TRIVENT
PUBLISHING

TRIVENT
Ethics | in Science & Technology

Heading Towards Humans Again

Aspects of Bioethics in the New Age of Science

Edited by

Miroslav Radenković

Applied Ethics: From Bioethics to Environmental Ethics, Vol. 4

Heading Towards Humans Again: Aspects of Bioethics in the New Age of Science
Ed. Miroslav Radenković
ISBN 978-615-6405-05-0
Applied Ethics: From Bioethics to Environmental Ethics, Volume 4
Series ed. János I. Tóth
Responsible publisher: Teodora C. Artimon
ISSN 2630-8185
https://trivent-publishing.eu/45-applied-ethics

First published in 2021 by Trivent Publishing

Trivent Publishing
1119 Budapest, Etele ut 59–61
Hungary

For more information, visit our website: http://trivent-publishing.eu

APPLIED ETHICS:
FROM BIOETHICS TO ENVIRONMENTAL ETHICS

Table of Contents

INTRODUCTION

Bioethics has become an important part of everyday life, not just including preclinical and clinical research, but also regular treatment of our patients. Still, there are numerous issues that we need to resolve each and every day, frequently going far beyond basic principles of autonomy, beneficence, nonmaleficence, or justice. This edited collection aims to challenge some critical cornerstones of today's contemporary bioethical concerns and interests.

The strong impression that bioethics, in spite of its normative quality, should take account of empirical data has been broadly acknowledged with cultural and social realities that are now more and more involved. To go further, it is clear that analytical methodologies of traditional bioethics could significantly advance by the use of empirical data regarding the intrinsic variety of human populations. In this context, we need to consider the technological revolution, nowadays reaching a whole new level. Fast changes that are not necessarily always positive, can lead to unavoidable discussions over posthumanism and transhumanism. Living in an increasingly technical world, do we need to redefine what is human? Well, we need to admit that all humans are merely mortals and death is an inevitable occurrence. In the world of artificial intelligence, biomedical technologies are facilitating the circumstances around natural death, primarily by sustaining human lives, thus making us more human after all.

If we go back to the basics, we know that in everyday clinical practice or research on humans, healthcare providers and investigators must be up to date with the current developments in bioethics. We can agree that, considering the utmost standards of keeping everyone's wellbeing, an informed consent still represents an ultimate anchor, always fastening the relationship between the doctor and their patient. And following all innovative technological advances, it is absolutely essential to adopt new multimedia communication strategies to further enhance comprehension of treatment-related or clinical trial-related information among patients and/or research participants. In this way, we need to be very careful considering prenatal genetic testing and screening, then gene and cell therapy including treatment of rare diseases, as well as the everlasting question regarding optional or mandatory vaccinations, for example against the human papillomavirus (HPV).

In everyday clinical practice, we are still struggling with bioethical challenges regarding chronic pain management, especially taking into account that pain may, not so uncommonly, last for months and years, and sometimes because of inadequate treatment. Pain is deeply subjective and should be recognized in our patients, and with a high bioethical sense, all the barriers to good chronic pain management should be surmounted. To continue, a specific and still controversial problem lies in the current medicalization of cannabis. Many argue that it is outdated and ultimately unethical to continue supporting the traditionally accepted position that there are no legitimate medical applications for cannabis, further suggesting that the use of cannabis should be addressed in a manner that avoids bias, presumption, or implicit judgment. On the other hand, the ethical implications of performance enhancing substances can still bring many substantiated and justified antidoping arguments.

Finally, it is important to underline that without educational and legislative components, bioethics as its stands today would simply not be complete. In that way, although we know that medicine-related bioethical issues are well defined, and many of those were adequately addressed, still there is an increasing necessity for guidelines in dental ethics in order to point at the ethical responsibility of everyone working in the dental practice towards patients, each other, as well as toward actions in dental research. Likewise, the role of ethics committees in medical faculties lay in providing the basic principles of professional ethics to healthcare providers, as well as in defining codes of conduct for all faculty employees, but also in providing opinions on controversial ethical issues that are of importance for conducting scientific, medical and public health research in healthcare institutions.

This edited collection, by tackling all these mentioned bioethical issues, will bring further views and concepts in the field, leading us to a better understanding of contemporary bioethics. I would like to express my deep gratitude to all the authors, reviewers, and to Trivent Publishing for providing unreserved support and help.

Prof. Dr. Miroslav Radenkovic
Faculty of Medicine, University of Belgrade

NEUROANTHROPOLOGICAL TURN IN BIOETHICS? INSIGHTS FROM INTERCULTURALITY

Attila Dobos[1]

INTRODUCTION

The idea that bioethics, despite its normative character, should include empirical data has been widely recognized in the last two decades and mostly subsumed under headings like "empirical turn," "cultural turn," "anthropological turn," "empirical bioethics," "social scientific critique of bioethics" or "culturally sensitive bioethics."[2] From the 1970s onwards, the works of genuine scholars such as Renee Fox or Arthur Kleinman have been grounded in this critical approach through fieldwork experiences in e.g., China, Zaire (now: the Democratic Republic of the Congo), India, Japan, Russia, etc. From one point of view, the empirical turn addressed the notion that the philosophical (Anglo-Saxon) analytic methodologies of traditional bioethics are habitually too abstract and insensitive of cultural and social realities, so often fail to properly understand and address the real issues thoroughly embedded in social life. One of the consequences was a broadening of the field of experience due to self-reflection, which is so central for the understanding of anthropological knowledge. In this matter, the empirical drew the anthropological, as fieldwork experiences revealed the vast variety of healing practices, health world views, and institutional

---

1 Semmelweis University, Hungary.

2 cf. K. Mbugua, "Respect for cultural diversity and the empirical turn in bioethics: a plea for caution." *Journal of Medical Ethics and History of Medicine* 5 (2012): 1-5.

approaches. To summarize: parallel medical universes appeared beside the universe of 'western' bioethics.

This insight–that empirical data should be used in bioethics–raises at least two issues in philosophy that need to be addressed:

- the problem of Eurocentrism in philosophy in general, and
- the question of the value-fact relationship in ethics.

Additionally, one might claim that these two domains are intertwined through the conception of ethics in different traditions around the world (e.g. the debate on Confucian vs. Kantian ethics). Since the production of knowledge is accumulating at a pace we have never seen before, it is of utmost importance to try to clarify the main concerns regarding its use in bioethics. In the following paragraphs, I contribute to this initiative.

EUROCENTRISM AND UNIVERSALISM

Since the Age of Discovery, it has been a struggle to accept the fact that inhabitants of distant places, far from Europe, are also human beings with equal merits. For instance, in his famous lecture in Vienna, in May of 1935, a leading figure of contemporary philosophy at that time, Edmund Husserl stated that in the broader sense of the Aristotelian *animal rationale* concept, even the Papua people are humans and not animals.[3] This remark was not meant to judge the status of the Papua, but the wording of sentences later on the same page (e.g. "…so wie der Mensch und Selbst der Papua…" [as Mankind and the Papua themselves too]) clearly shows the presumption that these people are far enough not to perceive them automatically and necessarily as humans.[4]

For the last 200 years of the European perspective, unconsciously or sometimes even consciously, all other conceptual traditions have been deemed non-philosophies, religions, or magical practices. At most, philosophy per se was defined as 'what the Greeks were doing.'

---

[3] "Nach der guten alten Definition ist der Mensch das vernünftige Lebewesen, und in diesem weiten Sinne ist auch der Papua Mensch und nicht Tier." cf. *Husserliana* VI. 2. Auflage, ed. Walter Biemel (Haag: Martinus Nijhoff, 1976), 337-338.

[4] The use of Papuans in this context was widespread in German philosophy that time, perhaps due to the fact that New Guinea was a former colony of Germany before the 1st World War. cf. T. Miettinen, *The Idea of Europe in Husserl's Phenomenology* (Helsinki: University of Helsinki, 2013), 146.

Conceptual frameworks and mental constructions other than those built upon the Greco-Roman tradition were by definition excluded from the domain of universal philosophy. Interestingly, this exclusive western conception of philosophy (that is rooted in the Greek tradition beginning with Thales) became widespread only after the end of the 18th century in the historiography of philosophy.[5] Before that, the great intellectual traditions were also valued as philosophies.

But why must we refer to only a particular activity as philosophy and other intellectual activities (e.g. what has been happening in other parts of the world) as something else? Why would this be problematic? In principle, it is because at the same time, philosophy is often conceived of as the highest form of intellectual activity and described as the 'mother of all knowledge.' It implies that other forms of intellectual traditions cannot apply for these honours. (Moreover, in the context of post-colonial self-reflection, a power dimension of this problem has also been unveiled: who owns the positions from which these definitions can be made and maintained? However, I do not intend to address this issue in this paper.)

It is the mission of the so-called intercultural philosophy movement to reflect on these constraints. Amongst many more, the works of Heinz Kimmerle, Franz Martin Wimmer, Ram Adhar Mall, Wim van Bimsbergen, Jan Hoogland, and Hamid Reza Yousefi are on point. Interestingly, most of the thinkers in this movement are affiliated with German associations or publishers: Gesellschaft für Interkulturelle Philosophie, Wiener Forum interkulturellen Philosophierens, Internationale Kolleg Morphomata, *Polylog: Zeitschrift für interkulturelles Philosophieren.* They all emphasize polylogue and many-voiced speech instead of universalist rhetoric, and pose radical challenges to the hegemony of Eurocentric philosophy. Firstly, their movement tends to rediscover and revaluate the great intellectual traditions of China and India, but it also opens up other regions, e.g., African philosophy. In doing so, they are holding a mirror to the Western mind. Secondly, intercultural philosophy offers support for the struggle with the difficulties of comparative philosophy, which could also be understood as the first stage towards a clearly intercultural philosophy. The main concerns about the (in)commensurability of different traditions that have

[5] cf. F.M. Wimmer, "How Are Histories of Non-Western Philosophies Relevant to Intercultural Philosophizing?" *Confluence* 3 (2015): 127-128.

developed independently in remoteness have created the domain of comparative philosophy. There are many levels of this debate, and it is clear that by no means could one find this task simple. Even if we claim that these traditions are commensurable, the ground for comparisons has to be transparent, as Bernard Waldenfels pointed out with the help of the *tertium comparationis* concept. "Comparing occurs on the horizon of a certain culture and not in a transcultural 'nowhere.' Temporal, spatial, and thematic horizons (which Nietzsche and Husserl discussed in detail) are borderlines of experience that, like our own shadows, walk with us."[6] This transparency requires humble self-reflection and self-critique at an advanced level–that is the very essence of anthropology. In this way, the empirical data from social sciences, including data about human beings radically different from ourselves, could contribute to the self-understanding of philosophy, even if within philosophy itself, the relation towards empirical data is highly problematic. Just to mention one well-known dispute some years ago made by Stephen Hawking in the introductory notes of his Grand Design book: "…but philosophy is dead. Philosophy has not kept up with modern developments in science, particularly physics."[7]

In the case of bioethics, this particular Eurocentric aspect of philosophy is enforced by the commonly conceived nature of medicine: it has to be just as universal as the human body. But exactly this character of the human body has been challenged recently. It has been claimed that the central focus of medicine should be on the continuous interdependence of the physical environment, the socio-cultural environment, and the body. The recent explosion of epigenetic research[8] has also supported this concept with a vast amount of research data. Following from all of these, along with the critique of Eurocentrism in philosophy, universalist claims have been questioned heavily.

[6] cf. Bernhard Waldenfels, "Comparing the incomparable: crossing intercultural borders," in *The Agon of interpretations*, ed. Ming Xie (Toronto: University of Toronto Press, 2014), 85.

[7] S. Hawking, L. Mlodinow, *The Grand Design* (New York: Bantam Books, 2010), 2.

[8] "The past 15 years have seen an explosion of discoveries related to the cellular regulation of phenotypes through epigenetic mechanisms." Nita Ahuja, Anup R. Sharma and Stephen B. Baylin, "Epigenetic Therapeutics: A New Weapon in the War Against Cancer," *Annual Review of Medicine* 67 (2016): 73.

THE VALUE – FACT QUESTION

Another problem with these "turns" in bioethics is methodological: how can methodologies of the socio-empirical sciences be integrated into the traditional philosophical methods of normative ethics?[9]

The problem of using empirical data in bioethics requires some further clarification and a bold outline of the epistemological background. Without bringing this discussion deep into meta-ethical dimensions, a few thoughts should be noted here concerning ethical naturalism to reveal the cornerstones of positions such as the one I propose in this argument. In the field of ethics, there has been a longstanding debate on the relation between normative statements (values) and positive statements (facts), for instance, on the claim that facts cannot justify values; values can only be justified by philosophical or conceptual inquiries about values. From the period of the Enlightenment, this view has been clearly expressed as Hume's law, or the 'is-ought problem". The naturalistic fallacy is another claim about the same problem. To put it simplistically: it means that one cannot make statements about what ought to be, based on statements about what is. The very same people who make these claims also claim that bioethics should not use positive statements created by science as long as it wants to be normative. It would mean that the discovery of human cognition and behaviour has no relevance for the development of bioethics. The inadequacy of this conclusion cannot be fully understood without understanding the inadequacy of the epistemological claims.

For a deeper understanding of empirical data usage in bioethics, one should differentiate between different types of data and their different uses. I will only refer here to Soren Holm's insightful defence of empirical bioethics.[10] He makes distinctions between biological, psychological, and sociological data and shows how moral theory and bioethics either implicitly rely on these data or should make use of them to be realistic and user-centred. Nevertheless, it is true that for proponents of the above-mentioned is-ought problem, who claim that

[9] See: Marcel Mertz, Jan Schildmann, "Beyond integrating social sciences: Reflecting on the place of life sciences in empirical bioethics methodologies," *Medicine, Health Care and Philosophy* 2 (2018): 207-214.

[10] See: *Engaging the World: The Use of Empirical Research in Bioethics and the Regulation of Biotechnology*, ed. Soren Holm, Monique F. Jonas (Washington, DC: IOS Press, 2004) 3-8.

"is" and "ought" are completely unrelated, psychological realities will not count in their arguments. Accordingly, those who believe that bioethicists should not turn their attention to certain problems only because they affect a great number of people, will not consider social scientific data relevant to their standpoints. In my view, however, bioethics that is based only on Eurocentric conceptual schemes and arguments will necessarily remain in the logical circle it painted itself into, trapped in universalist landscapes that were drawn by slender, monochrome pencils.

NEUROANTHROPOLOGY – THE NEW KEY?

By now, the psychological universality of mankind has been thoroughly investigated – not just at the behavioural level, but due to the revival of epigenetic research facilitated by the technological revolution, also at the cellular level. The results suggest a variability involving differences in environmental adaptation. Surprisingly, despite all of this investigation, the behavioural sciences database should be considered extremely narrow. An astonishing study[11] found that between 2003-2007, in top journals in six different subfields of psychology (including social psychology), 96% of studied subjects were from WEIRD (Western, Educated, Industrialized, Rich, Democratic) countries. Additionally, in most cases they were undergraduate students in the authors' psychology courses. For instance, in the *Journal of Personality and Social Psychology*, one of the leading journals in social psychology and thought to be the subfield of psychology most sensitive to the social context of the subjects, in total around 80% of the subjects were WEIRD undergraduate students, i.e. a very narrowly selected part of the global population. The use of these data in many cases is underpinned by the assumption "that either there is little variation across human populations, or that these 'standard subjects' are as representative of the species as any other population."[12] These data in behavioural sciences then are used to formulate knowledge about human beings in general. Obviously, this is not what one would expect from science. Heinrich and his colleagues have done a review of the comparative database from across the behavioural sciences and

---

[11] J. Henrich et al., "The weirdest people in the world?" *Behavioural and brain sciences* 33 (2010): 61–135.
[12] Ibidem, 61.

concluded that both of these assumptions are wrong: there is significant variability across different populations, and WEIRD subjects are amongst the least representative for making general claims about humanity.

Following from the above, I would like to suggest the possibility of using current scientific research data from the "fresh" field of neuroanthropology[13] as the ground for supporting further developments in bioethics. This epistemic field – which could be considered a subfield of psychological anthropology by integrating neuroscience and anthropology – originated in the '70s, but its scholarly and institutional recognition date only from 2007. The interdependence and correlation of the socio-cultural environment, the physical environment, and the human body lie in the central focus of this field. It has proven with extensive research that the differences in human perception, cognition, metaphorical thinking, and the "enculturation of the brain" are without any doubt reflected at the cellular level, too. The concept of personality and the ideas about the self have also been proved to be different in various parts of the globe.[14] But if so, how could we assume the proper functionality and legitimacy of abstract normative concepts, which are based on the supposedly known, also abstract and highly generalized human condition?

So why neuroanthropology? Because it seems that 1) the human body is standing in the *axis mundi*, where philosophical and scientific understanding is questioned, 2) the anthropological approach is necessary to reach the level of critical self-reflection providing possibilities for further development. Medicine, as well as bioethics, tends to be universal and based on the conception of the universal human body. Philosophy, as well as ethics, tends to be Eurocentric and based on the traditional Greco-Roman heritage. But just as anthropological data have intersected the boundaries of the Eurocentric circle in philosophy through the intercultural philosophical movement, neuroanthropology can convincingly show that the biological

---

[13] *The Encultured Brain,* ed. Daniel H. Lende, Greg Downey (Cambridge, MA: The MIT Press, 2012).

[14] R. A. Shweder, E. J. Bourne. "Does the Concept of the Person Vary Cross-Culturally?" in *Cultural Conceptions of Mental Health and Therapy. Culture, Illness, and Healing*, ed. Anthony J. Marsella, Geoffrey M. White (Dordrecht: Springer Netherlands, 1982), 97-137.

functioning of the human body could vary across different cultures and populations.

CONCLUSION

In this paper, I tried to argue for and support the claims that bioethics has to open up and acknowledge the importance of the use of empirical data on the intrinsic variety of human populations. In doing so, I exposed two background concepts that hinder this process: the Eurocentric tradition in philosophy and the value-fact debate in ethics. In my view, empirical data from anthropology can serve as a ground to see further and step out of the conventional circles of ideas. It can support the evolution of intercultural philosophy on the one hand, and give rise to neuroanthropology on the other. If we accept that anthropology is the field of study that institutionalized and professionalized the experience of strangeness[15], it seems we should also agree with Novalis: "Philosophy is really homesickness – the desire to be everywhere at home."[16] It may perhaps reveal the inherent interconnectedness of these two fields, or as Heinz Kimmerle states: there is a 'philosophical-historical necessity' for an intercultural dialogue within philosophy.[17]

ACKNOWLEDGEMENTS

This work is supported by the EFOP-3.6.1-16-2016-00022 project. The project is co-financed by the European Union and the European Social Fund.

REFERENCES

Ahuja, Nita, Anup R. Sharma and Stephen B. Baylin. "Epigenetic Therapeutics: A New Weapon in the War Against Cancer." *Annual Review of Medicine* 67 (2016): 73–89.

15 A. A. Teraoka, "Is Culture to Us What Text is to Anthropology?" *The German Quarterly* 62/2 (1989): 188.

16 Novalis, *Notes for a Romantic Encyclopaedia* (New York: State University of New York Press, 2007), XXV.

17 J. Hoogland, "The necessity of intercultural philosophy," in *Time and Temporality in Intercultural Perspective*, ed. D. Timersma and H.A.F. Oosterling (Amsterdam: Rodopi, 1996), 25.

Walter, Biemel, ed. *Husserliana VI. 2. Auflage*. Haag: Martinus Nijhoff, 1976.

Hawking, Stephen and Leonard Mlodinow. *The Grand Design*. New York: Bantam Books, 2010.

Hedgecoe, Adam M. "Critical bioethics: beyond the social science critique of applied ethics." *Bioethics* 18 (2004): 120-143.

Henrich, Joseph et al. "The weirdest people in the world?" *Behavioural and brain sciences* 33 (2010): 61-135.

Holm, Soren, and Monique F. Jonas, ed. *Engaging the World: The Use of Empirical Research in Bioethics and the Regulation of Biotechnology*. Washington, DC: IOS Press, 2004.

Kleinman, Arthur and L. Eisenberg, ed. *The Relevance of Social Science for Medicine*. Reidel Publishing, 1981.

Lakoff, Georg. "Mapping the Brain's Metaphor Circuitry: Metaphorical Thought In Everyday Reason." *Frontiers in Human Neuroscience* 8 (2014): 958.

Lende, Daniel H., Greg Downey, ed. *The Encultured Brain*. Cambridge, MA: The MIT Press, 2012.

Mbugua, Karori. "Respect for Cultural Diversity and the Empirical Turn in Bioethics: A Plea for Caution." *Journal of Medical Ethics and History of Medicine* 5 (2012): 1-5.

Mertz, Marcel, Jan Schildmann. "Beyond Integrating Social Sciences: Reflecting on the Place of Life Sciences in Empirical Bioethics Methodologies." *Medicine, Health Care and Philosophy* 2 (2018): 207-214.

Miettinen, Timo. *The Idea of Europe in Husserl's Phenomenology*. Helsinki: University of Helsinki, 2013.

Novalis. *Notes for a Romantic Encyclopaedia*. New York: State University of New York Press, 2007.

Shweder, Richard A., Edmund J Bourne. "Does the Concept of the Person Vary Cross-Culturally?" In *Cultural Conceptions of Mental Health and Therapy. Culture, Illness, and Healing*, ed. Anthony J. Marsella, Geoffrey M. White, 97-137. Dordrecht: Springer Netherlands, 1982.

Teraoka, Arlene Akiko. "Is Culture to Us What Text is to Anthropology?" *The German Quarterly* 62(2) (1989): 188.

Timersma, Douwe and Henk Oosterling, ed. *Time and Temporality in Intercultural Perspective*. Amsterdam: Rodopi, 1996.

Leigh, Turner. "Anthropological and Sociological Critiques of Bioethics." *Bioethical Inquiry* 6 (2009): 83-98.

Waldenfels, Bernhard. "Comparing the Incomparable: Crossing Intercultural Borders." In *The Agon of interpretations*, ed. Ming Xie, 83-98. Toronto: University of Toronto Press, 2014.

Wimmer, Franz M. "How Are Histories of Non-Western Philosophies Relevant to Intercultural Philosophizing?" *Confluence* 3 (2015): 127-128.

TRANSHUMANISM OR PRAGMATISM?

Alexander Kremer[1]

INTRODUCTION

Since the permanent scientific and technological revolution reached a new level, I am convinced that we have to analyse it in a more general context, not only on the level of technology. New technologies, which not only improve our given skills but also create new abilities that can change 'the future of humanity thoroughly, started to be invented with the beginning of the third millennium. These changes are not necessarily positive, but they have roused the passion of many people. This enthusiasm emerges mostly without theoretical support (in the case of laypeople, criminals, military leaders, etc.), but we can also find the theoretical background in the forms of posthumanism and transhumanism.

Posthumanism is a loose philosophical movement, which means those branches of philosophers who try to explain humanity's future in an increasingly technical world. Posthumanism has five main types: a) antihumanism, b) cultural posthumanism, c) philosophical posthumanism, d) posthuman condition, and e) transhumanism. Transhumanism is the most radical part of posthumanism, which tries to redefine what is human. The reason for the birth of posthumanism is the increasing development of AI and different ETs. Scientists are creating more and more human-like AI, and AI is becoming similar to humans every year. (Look at e.g., Sophia, the first android robot who received Saudi Arabian citizenship, gives many interviews and sings a song in The Tonight Late Night Show. Nevertheless, Sophia is only one of the

[1] University of Szeged, Hungary.

"independently" communicating and learning robots today.) What are ETs? Are they good or bad from a moral point of view? The domain of ETs is extremely broad, and it includes very diverse territories from cosmetic surgery to sports and cognitive enhancements, to genetic engineering, and further.

Transhumanism prefers the perspective in which humans and robots are directly connected, and human features are equal to technological features (see, for example Neil Harbisson, the first officially accepted cyborg artist and transspecies activist based in New York, or the transhumanist Nick Bostrom, cited in this paper). It follows from this that transhumanists want not only to transcend the traditional humanist concept of the human being but also to destroy it. They argue that we should not preserve the human being if we can become something better.

ARGUMENTATION

Nevertheless, we have to see clearly the decisive question and not let ourselves be fascinated by the new scientific and technological potentialities. The decisive questions are namely not technological, but rather moral and political questions. To understand this standpoint, we should start with the clear fact that *natural processes are morally neutral in themselves.* The moral character of the natural processes depends on the moral content of the aims and activities at and in which we use them (e.g. the nuclear chain reaction functions not only in a nuclear bomb but also in a nuclear power plant). If we use a neutral natural process for a bad aim, the neutral natural process seems morally wrong because it is embedded in a morally bad context. If we use a neutral natural process for a morally good aim, this process seems to be morally good as well, because it is embedded in a good context of social activity. However, we should not forget that the natural process remains neutral; only the human aim and activity can be morally bad or good.

Let us use an *analogy*, and we can recognize that it is always true in connection with the latest scientific and technological developments, as well. *New technologies function as natural processes.* These are morally neutral in themselves, and from a moral point of view, everything depends on their social context (e.g. CRISPR-Cas9 technology is beneficial for medical treatments, but we can also create chimeras. If we use new technology *as therapeutic technology*, it is morally good. If we use new technology *as enhancement technology* – as unnecessary augmentation of a human capacity or bodily form –, it is bad.) Everything depends on the moral content of the human aim and activity in which the new technology is embedded. The latter one remains neutral, but the human context (the human aim and activity) can be morally good or bad.

At this point, we should not forget a significant *difference*. Humanity could *not* create such technologies in earlier centuries, which had such long-running consequences. But nowadays we can change the human race if we change, for example, the human genome with the CRISPR-Cas9 genetic engineering technology. As we already cannot control global warming and the contamination of our environment, etc., we will not be able to do it regarding the modified human genome or artificial intelligence. *It follows from this that it is our moral obligation and responsibility towards ourselves and the future human generations* to move forward *very slowly and cautiously* regarding these dangerous new technologies, which can drastically change the human race and our environment.

However, the moral decision is not enough. We need the power to realize the moral decision, and this power can be found only in politics. Human communities can validate their moral decisions exclusively in the proper legal and political decisions that are results of democratic social debates. These considerations significantly increase the importance of political decisions (together with that of the political systems and institutions), especially the importance of democracy. It follows from this that in our epoch, political decisions are much more important than scientific questions. This recognition is that is missing in transhumanism and can be found in pragmatism.

TRANSHUMANISM

After placing this new social tendency, the always increasing scientific and technological development, into a broader moral and political context, it becomes clear why most transhumanist's standpoints are problematic. Most representatives of transhumanism prefer individualism and individual freedom, and do not mention moral and political questions. They are mostly playing with the technological novelties in the sandbox of the scientific and technological revolution in a complacent and irresponsible way, and do not deal with political institutions and democracy. If we look at e.g. Nick Bostrom's texts, then we can find only individualism and negligence regarding the moral and political problems:

> To start with, *transhumanists typically place great emphasis on individual freedom and individual choice, especially when it comes to ETs.* The reason for this is twofold.
>
> First, it is a fact that *humans differ widely in their conceptions of what their own perfection would consist in.* Some want to develop in one direction, others in different directions, and some prefer to stay pretty much the way they are (whether because of religious or other motives). It would neither be feasible nor desirable to impose one common standard that we should all aspire to. *The best approach then is to let people choose for themselves* which ETs they want to use on themselves if any.
>
> The second reason for this element of individualism is the poor track record of collective *decision-making in the domain of human improvement.* The eugenics movement, for example, is thoroughly discredited; and other collectivist utopian projects have mostly been total failures.[2]

We can also quote from a different book, *Post- and Transhumanism. An Introduction.*[3] If we look at the chapter about politics (written by James Hughes), we can find the next individualistic approach:

> While transhumanists have more normative consensus than the posthumanists, they are not in agreement about very much beyond their core commitments to non-anthropocentric personhood, techno-optimism, and individual liberty. *Transhumanists generally agree that the development of life extension and cognitive enhancement therapies should be pursued and that individuals should be able to choose to use them. They agree that individuals should be able to use reproductive technologies to have children, and to choose those children's characteristics.*[4]

[2] Nick Bostrom, *Transhumanist Values* (2001), www.nickbostrom.com/tra/values.html (Last accessed: October 13th, 2019).

[3] Robert Ranisch and Stefan Lorenz Sorgner, eds., *Post- and Transhumanism. An Introduction* (Frankfurt am Main: Peter Lang GmbH, Internationaler Verlag der Wissenschaften, 2014).

[4] James Hughes, "Politics," in *Post- and Transhumanism. An Introduction*, ed. Robert Ranisch and Stefan Lorenz Sorgner (Frankfurt am Main: Peter Lang GmbH, Internationaler Verlag der Wissenschaften, 2014, p. 139).

If we read the concluding passages of the chapter on politics, we can read the next characteristic sentences:

> In the end, transhumanism and posthumanism are distinguished by the work they are trying to do. *Transhumanists want to create a future in which every individual can use science and technology to free themselves from the constraints of nature and fulfill their own concepts of the good life.* Posthumanists are attempting to show how our concepts of the body, nature, happiness, freedom, and fulfillment are socially constructed so that we are pushed to a deeper questioning about the future that we create. The transhumanist project grounds the posthumanists in applied biopolitics defending our bodies, brains, and reproduction from counter-Enlightenment. Posthumanists are the skeptical intellectuals teasing apart the ways that the revolutionary project needs to question its own categories and agendas so that the revolution will be truly liberatory. By engaging with one another the two communities have much to gain.[5]

Let us have a look at the chapter "Morality" (by Robert Ranisch), and we can read the next passage:

> *There is no comprehensive transhumanist morality or moral theory.* It is still possible to identify common moral claims concerning right or wrong action, what is of value and what constitutes a good or virtuous character. *Without a doubt, transhumanism as a techno-optimistic political and cultural movement is distinguished by a specific set of normative assumptions, and transhumanist organizations dedicate themselves to moral questions.* Nevertheless, transhumanism comes in many forms, representing a wide spectrum of moral and political orientations.[6]

[5] James Hughes, "Politics," in Post- and Transhumanism. An Introduction, ed. Robert Ranisch and Stefan Lorenz Sorgner (Frankfurt am Main: Peter Lang GmbH, Internationaler Verlag der Wissenschaften, 2014, p. 145).

[6] Robert Ranisch, "Morality," in *Post- and Transhumanism. An Introduction*, ed. Robert Ranisch and Stefan Lorenz Sorgner (Frankfurt am Main: Peter Lang GmbH, Internationaler Verlag der Wissenschaften, 2014), 149.

"Most transhumanists "explicitly affirm the possibility to overcome human biological nature in a radical way. To realize transhumanist visions, "ideally, everybody should have the opportunity to become posthuman."[7] By the same token, technology is not yet developed, and admittedly, "there is no manner by which any human can become a posthuman"[8] *at the moment.* As a consequence, *everything transhumanists can hope for is that people are willing to submit themselves to the transhumanist "social experiment,"*[9] which makes them part of an evolutionary chain that may eventually lead to posthumanity. *By contrast, achieving posthumanity is not the aim of bioliberals. While a few bioliberals embrace such possibility,*[10] *this should rather be seen as a side effect of human enhancement.* Even though "there is nothing wrong" that we may end up in posthumanity, "becoming transhumans is not the agenda."[11]

CONCLUSION: PRAGMATISM

Contrary to many transhumanists, *pragmatists support democracy.*[12] Let us take either classical pragmatists or neopragmatists. That is why I am convinced that we do not have to destroy the human world, taken in a humanist sense. I believe that we have an alternative option in pragmatist philosophy. Pragmatism was (and is) also a very loose philosophical

---

[7] N. Bostrom, "Transhumanist Values," *Journal of Philosophical Research* 30 (Supplement - Ethical Issues for the Twenty-First Century) (2005): 10.

[8] Humanity+ (n.d.), *Transhumanist FAQ: 3.0.* http://humanityplus.org/philosophy/trans humanist-faq/ (accessed December 3, 2019).

[9] M. Walker, "Ship of Fools: Why Transhumanism Is the Best Bet to Prevent the Extinction of Civilization," in *H±: Transhumanism and Its Critics*, ed. Hansell, G.R. and Grassie, W. (Philadelphia: Metanexus Institute, 2011), 94-111.

[10] J. Savulescu, J. "The Human Prejudice and the Moral Status of Enhanced Beings: What Do We Owe the Gods?" in *Human Enhancement*, ed. Savulescu J./ Bostrom N. (Oxford: Oxford University Press, 2011), 211-47.

[11] J. Harris, *Enhancing Evolution: The Ethical Case for Making Better People* (Princeton University Press, 2007), 39

[12] I am convinced that many transhumanists (e. g. Nick Bostrom) support the extreme individualism (see N. Bostrom's text mentioned above) that cannot be accepted even in a liberal democracy since its vocabulary must be shared. It means that its citizens should use self-limitation. My view is in harmony with J. St. Mill's opinion: "That the only purpose for which power can be rightfully exercised over any member of a civilized community, against his will, is to prevent harm to others." (J. Stuart Mill, *Utilitarianism and On Liberty*Second edition. Ed. with an Introduction by Mary Warnock (Blackwell Publishing, 2003), 94.

movement, but its participants have common features like the practice-oriented interpretation of human life and truth, naturalism, meliorism, "experientialism" and experimentalism, and what is the most important for us: democracy. *If we look at John Dewey's, Richard Rorty's, and Richard Shusterman's philosophy, it is clear as day that pragmatism tries to maintain democracy and humanism.*

It is already commonplace that we cannot prove any ultimate metaphysical essence of the human being. However, already the classical pragmatist's naturalism meant that human being is a transactional being since it always lives not only as a part of nature but also in mutual practical relationships with nature. Humans form a biological race, and we cannot live without natural goods as long as we remain humans. It follows from this that we are (and will be) a part of nature and we are not only intelligent but also emotional and suffering beings. As Dewey put it, we are "transactional beings," but democracy is the best way to develop our human features, since – according to his famous expression – "democracy is a way of life."[13]

Richard Rorty replaced experience with language in his philosophy, but even he spoke about a new type of solidarity and democracy. Since humans do not have any intrinsic spiritual features that could be the basis of solidarity, he chooses a new basis. This new basis is the recognition that pain and suffering is a common feature of these rational beings who we are, and we have to increase our empathy. We have to make ourselves more and more sensitive regarding the suffering of other human beings, and this is the empirical basis of Rorty's new solidarity. Relying on this new solidarity, we can build a new type of human, the so-called liberal ironist who can create, in Rorty's opinion, a new liberal democracy, where politics is much more important than science.

Richard Shusterman replaced language not only with experience but also with the living soma. In his project, *somaesthetics* the soma is the "tool of tools" since we could not appropriate anything and create anything without our soma. (See perception, eating, drinking, artworks, etc.) On this basis, Shusterman prefers the emancipatory "somapower" to Foucault's oppressive, exploitative "biopower." Shusterman does not want to deal with political questions directly, but he supports democratic

[13] John Dewey, "Creative Democracy – The Task Before Us," in *The Essential Dewey (Volume I): Pragmatism, education, democracy*, ed. Larry A. Hickman and Thomas M. Alexander (Indiana University Press, 1998), 341.

social tendencies with his lectures, papers, books, and his somaesthetic practice.

If I want to summarize my standpoint, I can say that *pragmatists*, contrary to many transhumanists, support democracy and *our moral obligation and responsibility towards ourselves and the future human generations.* Thus, I am convinced that we are better off if we choose pragmatism!

REFERENCES

Bostrom, N. "Transhumanist Values." *Journal of Philosophical Research* 30 (Issue Supplement - Ethical Issues for the Twenty-First Century) (2005): 3-14.

—. *Transhumanist Values.* www.nickbostrom.com/tra/values.html (2001). Accessed: October 13th, 2019.

Dewey, John. "Creative Democracy – The Task Before Us." In *The Essential Dewey (Volume I): Pragmatism, education, democracy*, ed. Larry A. Hickman and Thomas M. Alexander. Indiana University, 1998.

Harris, J. *Enhancing Evolution: The Ethical Case for Making Better People.* Princeton University Press, 2007.

Humanity+ (n.d.): Transhumanist FAQ: 3.0. http://humanityplus.org/philosophy/transhumanist-faq/. Accessed December 3, 2013.

Mill, J. Stuart. *Utilitarianism and On Liberty.* Second editon. Ed. with an Introduction by Mary Warnock. Blackwell Publishing, 2003.

Ranisch, Robert and Stefan Lorenz Sorgner, ed. *Post- and Transhumanism. An Introduction.* Frankfurt am Main: Peter Lang, 2014.

Rorty, Richard. *Contingency, Irony and Solidarity.* Cambridge: Cambridge University Press, 1989.

Savulescu, J. "The Human Prejudice and the Moral Status of Enhanced Beings: What Do We Owe the Gods?" In *Human Enhancement,* ed. Savulescu J. and Bostrom N., 211-247. Oxford: Oxford University Press, 2009.

Shusterman, R. *Pragmatist Aesthetics, Living Beauty, Rethinking Art,* 2nd ed., New York: Rowman and Littlefield, 2000.

—. *Body Consciousness.* Cambridge, New York: Cambridge University Press, 2008.

—. *Thinking Through the Body. Essays in Somaesthetics.* New York: Cambridge University Press, 2012.

Walker, M. "Ship of Fools: Why Transhumanism Is the Best Bet to Prevent the Extinction of Civilization." In *H±: Transhumanism and Its*

Critics, ed. Hansell, G.R. and Grassie, W., 94-111. Philadelphia: Metanexus Institute, 2011.

Warnock, Mary, ed. *Utilitarianism and On Liberty*. Oxford: Blackwell Publishing Ltd, 2003.

THE DANGER OF FALSE PROMISES AND HOPES IN THE NEW AGE OF SCIENCE: THE CASE OF GENE AND CELL THERAPY

Oana Iftime,[1] Ștefana-Maria Petruț[1]

INTRODUCTION

The field of modern medical biotechnologies developed rapidly in terms of gaining attention and support, as well as in producing scientific literature and generating ethical debates. Also, a lot of valuable applications were established. Still, many medical tools that were initially acclaimed as most promising seem to have trouble in finding their way into the real world. Gene and cell therapy represent some of the most prominent examples. Decades of investments and efforts have resulted in few clinical applications, some of them ephemeral, so that in spite of the enthusiastic discourses still present in the public and academic media we got to a point where the "promises and hurdles" of these fields might seem totally imbalanced.

GENE AND CELL THERAPY IN THE REAL WORLD

During the last few decades, therapeutic gene manipulation has been suggested for and attempted to various extents in cancer, several types of immunodeficiencies, Alzheimer, Parkinson and Huntington diseases, muscular spinal atrophy, cystic fibrosis, retinopathies, as well as diabetes, cardiac failure, and others. Stem-cell therapy was also proposed for a

---

[1] University of Bucharest (Department of Genetics, Faculty of Biology), Romania.

wide range of conditions, many of them common with gene therapy targets.

In gene therapy, nucleic acids are introduced into the organism in order to prevent or treat a disease. They enter the cells by means of a carrier that is either a vector derived from a virus (lentivirus, poxvirus, adenovirus, adeno-associated virus, retrovirus, human foamy virus, herpes virus a. o.) or one of the molecular complexes or polymers proposed for use in gene therapy (e. g. liposomes, chitosan derivatives, poly(L-lysine), poly(ethylenimine) derivatives a. o.).[2] In order to obtain significant results, a substantial distribution of the delivered gene and persistence of its functioning should be ensured, which is very difficult as many gene carriers do not travel adequately through tissues and are subsequently lost. Also, they are not welcomed by the immune system, nor are the transgene or its product. Another significant problem is the risk of insertional mutagenesis posed by certain classes of carriers. The highly risky side of gene therapy research was tragically illustrated by the death of Jesse Gelsinger (1999) due to a violent systemic inflammatory response to the administered vector,[3] and the development of leukemia in four out of nine children treated for SCID (2003).[4] Improvements in vector design led to some better results,[5] but the difficulties are far from being overcame.

It was said that gene therapy had traversed a "hype cycle," comprising the over-enthusiasm elicited by its potential, the despair generated by its serious side-effects and the realism attached to improvements that might lead to success; it was also said that it might get again into another comparable sequence of events.[6] Indeed, the development of gene editing techniques stormed the enthusiasm of researchers. However, the supposedly revolutionary DNA modifying CRISPR-Cas 9 system and its

[2] Y. K. Sung and S. W. Kim, "Recent advances in the development of gene delivery systems," *Biomaterials Research* 12 (2019): 23-28.
[3] S. E. Raper et al., "Fatal systemic inflammatory response syndrome in a ornithine transcarbamylase deficient patient following adenoviral gene transfer," *Molecular Genetics Metabolism* 80 (2003): 148–158.
[4] M. P. McCormack and T. H. Rabbitts, "Activation of the T-cell oncogene LMO2 after gene therapy for X-linked severe combined immunodeficiency," *The New England Journal of Medicine* 350 (2004): 913–922.
[5] C. E. Dunbar et al., "Gene therapy comes of age," *Science* 359 (2018): 1-10.
[6] W. F. Kaemmerer, "How will the field of gene therapy survive its success?" *Bioengineering & Translational Medicine* 3 (2018): 166–177.

derivatives also pose lots of problems. The drawbacks of Cas 9 employment are significant – difficulties related to delivery into the cell, low efficiency of the enzyme itself and of the cellular repair processes,[7] toxicity mediated via a p53 mechanism,[8] and the completely undesirable capacity of the enzyme to induce unintended modifications in the DNA not only by off-target effects but also through its on-target activity.[9] Other tools considered promising, the CRISPR-Cas derived cytosine base editors (CBEs) were shown to induce a high rate of off-target modifications in mouse models,[10] as well as off-target DNA edits and extensive transcriptome-wide RNA cytosine deamination in human cells, generating missense, nonsense, splice site, 5' UTR, and 3' UTR mutations, effects that their discoverers tried to alleviate by further engineering of the CBE.[11] In short, all the examined molecular editors generate unintended, potentially dangerous modifications into the DNA, and other side effects.

Stem cells are able to differentiate into various types of cells and, for therapeutic purposes, they are supposed to be transplanted into the diseased organism. The undesirable aspects of their employment are linked to immune rejection, the risk of tumor development due to incomplete reprogramming in the cell population, difficulties in tissue targeting, retention and functional integration.[12] Human embryonic stem cells are also subject to intense ethical debates, as their obtainment requires the destruction of the embryos. The development of induced pluripotent stem cells (iPS), i.e. cells from adult sources that are re-programmed into an embryonic-like state, was regarded as a significant

[7] W. J. Dai et al., "CRISPR-Cas9 for *in vivo* gene therapy: promise and hurdles," *Molecular Therapy—Nucleic Acids* 5 (2016): 1-4.

[8] R. J. Ihry et al., "p53 inhibits CRISPR–Cas9 engineering in human pluripotent stem cells," *Nature Medicine* 24 (2018): 939–946.

[9] M. Kosicki, K. Tomberg, and A. Bradley. "Repair of double-strand breaks induced by CRISPR–Cas9 leads to large deletions and complex rearrangements," *Nature Biotechnology* 36 (2018): 765-771.

[10] E. Zuo et al., "Cytosine base editor generates substantial off-target single-nucleotide variants in mouse embryos," *Science* 364 (2019): 289-292.

[11] J. Grünewald et al., "Transcriptome-wide off-target RNA editing induced by CRISPR-guided DNA base editors," *Nature* 568 (2019): 433-453.

[12] G. Cossu et al., "Lancet Commission: Stem cells and regenerative medicine," *Lancet* 391 (2018): 883-910.

step in the field[13] and elicited over-enthusiasm, unfortunately not supported by the subsequent results.

Ex-vivo gene therapy that combines gene and cell therapy by using autologous genetically modified cells might represent a better approach and seems to have been successfully applied in treating some maladies.

The first gene therapy product ever marketed was Gendicine®, a vector bearing the gene encoding for the p53 tumour-suppressor. It received the Chinese authorities' endorsement in 2003 and it is, together with Oncorine®, another Chinese gene therapy drug, pretty controversial.[14]

In the USA, the Federal Drug Agency (FDA) has a current (as of 2019) number of seventeen "cellular and gene therapy products" approved.[15]

About half of them are cord blood preparations containing hematopoietic progenitor cells, a "classic" in cell therapy, further improved. As for the other products, MACI® (2016) is a scaffold type product bearing autologous cells, intended for the repair of cartilage defects of the knee[16]; GINTUIT® is another celularized scaffold product, designed for topical oral use[17]; PROVENGE® (2010) is to be prepared with autologous $CD54^+$ cells, activated, being advised as immunotherapy for metastatic prostate cancer[18]; LAVIV® (2018) is an autologous cell preparation advised for nasolabial fold wrinkles.[19] Two products make use of genetically modified cells – KYMRIAH® (2017) is to be prepared from CD-19-directed genetically modified autologous T-cells and is

---

[13] H. Inoue et al., "iPS cells: a game changer for future medicine." *The EMBO Journal* 33 (2014): 409-417.

[14] G. Ma et al., "Gene medicine for cancer treatment: commercially available medicine and accumulated clinical data in China". Drug design, Development and Therapy, 2 (2009): 115–122.

[15] FDA. *Approved Cellular and Gene Therapy Products* https://www.fda.gov/vaccines-blood-biologics/cellular-gene-therapy-products/approved-cellular-and-gene-therapy-products (accessed September 29, 2019).

[16] FDA. *Package Insert, MACI®* www.fda.gov/media/101914/download (accessed September 29, 2019).

[17] FDA. *Package Insert GINTUIT*, www.fda.gov/media/83264/download (accessed September 29, 2019).

[18] FDA. *Package Insert and Patient Information PROVENGE®*, www.fda.gov/media/78511/download (accessed September 29, 2019).

[19] FDA. *Package Insert and Patient Information Sheet, LAVIV®* www.fda.gov/media/80838/download (accessed September 29, 2019).

recommended for the treatment of certain types of leukemia and lymphoma, refractory or relapsed[20]; YESCARTA® (2017) is also an autologous genetically modified T cells-based therapeutic product for lymphoma.[21] Three other products consist in gene therapies: IMLYGIC® (initially approved in 2015) contains a genetically modified virus used for local therapy in recurrent melanoma[22]; LUXTURNA® (2017) is a gene therapy product employing an adeno-associated virus vector designed for the treatment of retinal dystrophy due to a biallelic RPE65 mutation, bearing a warning for serious adverse effects such as cataract, retinal tear, and maculopathy[23]; ZOLGENSMA® (2019) uses an adeno-associated virus vector for the treatment of spinal muscular atrophy with bi-allelic mutations in the SMN1 gene and it bears a warning of acute serious liver injury as well as the mention that its safety and effectiveness in the case of repeat administration have not been evaluated.[24]

In Europe, medicines based on genes, tissue or cells are named "advanced therapy medicinal products (AMTPs)" and have to be approved by The European Medicines Agency (EMA). So far, the EMA has approved ten products, out of which four have already been withdrawn. The cell therapies in use comprise Alofisel® (2018), that contains allogeneic mesenchymal adult stem cells and is recommended for perianal fistulas[25]; Spherox® (2017), a preparation of expanded autologous chondrocytes and self-synthesized extracellular matrix advised for articular cartilage defects[26]; Holoclar® (2015), employed in ocular burns, that contains autologous corneal epithelial cells expanded

[20] FDA. *Package Insert, KYMRIAH™* https://www.fda.gov/media/107296/download (accessed September 29, 2019).

[21] FDA. *Package Insert YESCARTA™* https://www.fda.gov/media/108377/download (accessed September 29, 2019).

[22] FDA. *Package Insert IMLYGIC* https://www.fda.gov/media/94129/download (accessed September 29, 2019).

[23] FDA. *Package insert LUXTURNA* https://www.fda.gov/vaccines-blood-biologics/cellular-gene-therapy-products/luxturna (accessed September 29, 2019).

[24] FDA. *Package Insert ZOLGENSMA®* https://www.fda.gov/media/126109/download (accessed September 29, 2019).

[25] EMA. *Product information Alofisel* https://www.ema.europa.eu/en/documents/product-information/alofisel-epar-product-information_en.pdf (accessed September 29, 2019).

[26] EMA. *Product information Spherox* www.ema.europa.eu/en/documents/product-information/spherox-epar-product-information_en.pdf (accessed September 29, 2019).

ex vivo[27]. Genetically modified preparations in use include Strimvelis®, Zalmoxis® and Imlygic®. Strimvelis (2016) is a preparation of autologous $CD34^+$ cells modified with a retroviral vector containing the human ADA cDNA sequence designed for the treatment of the severe combined immunodeficiency ADA-SCID; it exhibits as adverse effects a series of autoimmune disorders.[28] Zalmoxis (2016) is prepared from allogeneic T cells genetically modified with a retroviral vector encoding for two genes of interest and is recommended as adjunctive treatment in haploidentical hematopoietic stem cell transplantation in patients with blood cancer. Zalmoxis has notable adverse effects such as neoplasm, immune system disorders, and hepatic failure.[29] Glybera® (alipogene tiparvovec), known as the most expensive drug in the world, left the market in 2017, after five years from the initial approval. Glybera was a complex product, containing a human gene (LPL, encoding for lipoprotein lipase), packed into a chimeric vector derived from adeno-associated virus serotype 1, bearing the cytomegalovirus (CMV) promoter, a hepatitis virus posttranscriptional regulatory element and some adeno-associated virus serotype 2 derived inverted terminal repeat.[30] ChondroCelect® (2009), an autologous cell preparation advised for cartilage repair, is another product that is no longer authorized, as requested by the marketing authorisation holder.[31] So is Provenge®[32], while the authorisation of Maci expired following the holder's decision not to pursue its renewal.[33]

Some oligonucleotidic preparations were also approved either by the FDA or by the FDA and the EMA, but they had a complicated market route and/or significant side effects, which made them not very attractive, with the possible exception of Spinraza®, developed for spinal

---

27 EMA. *Product information Holoclar*, www.ema.europa.eu/en/documents/product-information/holoclar-epar-product-information_en.pdf (accessed September 29, 2019).
28 EMA. *Product information Strimvelis*, www.ema.europa.eu/en/documents/product-information/strimvelis-epar-product-information_en.pdf (accessed Sept. 29, 2019).
29 EMA. *Product information Zalmoxis*, www.ema.europa.eu/en/documents/product-information/zalmoxis-epar-product-information_en.pdf (accessed September 29, 2019).
30 EMA. *Product information Glybera*, www.ema.europa.eu/en/documents/product-information/glybera-epar-product-information_en.pdf (accessed September 29, 2019).
31 EMA. *ChondroCelect*, www.ema.europa.eu/en/medicines/human/EPAR/chondro celect (accessed September 29, 2019).
32 EMA. *Provenge*, www.ema.europa.eu/en/medicines/human/EPAR/provenge),
33 EMA. *Maci*, www.ema.europa.eu/en/medicines/human/EPAR/maci (accessed September 29, 2019).

muscular atrophy.[34] Other products were approved in the Philippines (Rexin-G®, an antitumoral), and Russia (Neovasculgen®, for arterial diseases).[35]

THE DANGER OF FALSE PROMISES AND HOPES

The biological hurdles in gene and cell therapies were presented above on the short. Also, a series of practical issues interfere with cell and gene therapy products manufacturing – their development is difficult, requiring an unusual trial design, the production process and the quality control are highly demanding, regulations are complex and may vary across countries, and some developers, as well as the insurance companies, feel that reimbursement might represent a serious challenge.[36]

Unfortunately, over-enthusiasm led to less accuracy in experimentation so that now, as some authors put it, "The literature is replete with instances of therapeutic interventions pursued on the basis of expert opinion and patient acceptance that ultimately proved ineffective or harmful when studied in well-controlled trials comparing them with the standard of care."[37] And, in spite of all evidence and signaled dangers, exaggerated optimism persists. In a comprehensive review from 2015, Trounson and McDonald expressed their confidence that there will be "many stem cell products that will meet the criteria for registered products in the established regulatory systems over the next 5 years"[38] – even though the paper listed lots of running experiments, trials

[34] C. A. Stein and D. Castanotto, "FDA-Approved oligonucleotide therapies in 2017," *Molecular Therapy* 25 (2017): 1069-1075.

[35] R. Goswami et al., "Gene therapy leaves a vicious cycle," *Frontiers in Oncology* 9 (2019): 1-25.

[36] W. F. Kaemmerer, "How will the field of gene therapy survive its success?" *Bioengineering & Translational Medicine* 3 (2018): 166–177. M. T. R. ten Ham et al., "Challenges in advanced therapy medicinal product development: A survey among companies in Europe, *Molecular Therapy: Methods & Clinical Development* 11 (2018): 121-130. J. F. Barlow, M. Yang, and J. R. Teagarden, "Are payers ready, willing, and able to provide access to new durable gene therapies?" *Value in Health*; 22 (2019): 642–647. J. Spoors, E. Croft, and A. Walker, "Regeneration X: The payer perspective on gene therapy," *British Journal of Healthcare Management*, 23 (2017): 158-166.

[37] P. W. Marks, C. M. Witten, and R. M. Califf, "Clarifying stem-cell therapy's benefits and risks," *The New England Journal of Medicine* 376 (2017): 1007-1009.

[38] A. Trounson and C. McDonald, "Stem Cell Therapies in Clinical Trials: Progress and Challenges," *Cell Stem Cell* 17 (2015): 11-22.

involving only a handful of subjects from which even less experienced a positive outcome, not always stable, and other not too convincing examples of progress in stem cells research. Anyway, the fifth year has begun, and their optimism had not proven to be substantiated yet, whilst the only truly reliable application of cell therapy continues to be the long-established hematopoietic stem cell transplantation.

Inflated optimism regarding cell and gene therapies seems to be shared by some officials: "We anticipate that by 2020 we will be receiving more than 200 INDs[39] per year, building upon our total of more than 800 active cell-based or directly administered gene therapy INDs currently on file with the FDA. And by 2025, we predict that the FDA will be approving 10 to 20 cell and gene therapy products a year based on an assessment of the current pipeline and the clinical success rates of these products."[40] What does such a positive estimate rely upon? Around 2,900 gene therapy clinical trials were registered from 1989 till 2018.[41] Still, as shown above, only a few products were approved for actual therapeutic use, with some of them rapidly leaving the market.

It is true that the industry has recently known a serious infusion of capital. Data collected by the Alliance for Regenerative Medicine show that in 2018 there were "906 Regenerative Medicine Companies worldwide, including Gene and Cell Therapies, and Tissue Engineering Therapeutic Developers" – 484 based in North America, 241 in Europe and Israel, 142 in Asia and the remaining in Oceania and South America, with the total level of financing in the industry reaching 13.3 billion US dollars in 2018 (73% increase from 2017), from which 9.7 billion were invested in gene-based therapies (64% increase from the previous year).[42]

---

[39] Investigational New Drug – the item described as such and also the procedure by which FDA allows an experimental drug to be shipped from state to state and used in clinical trials. See FDA. *Investigational New Drug (IND) Application* www.fda.gov/drugs /types-applications/investigational-new-drug-ind-application (accessed Oct. 8, 2019).

[40] FDA, *Statement from FDA Commissioner Scott Gottlieb, M.D. and Peter Marks, M.D., Ph.D., Director of the Center for Biologics Evaluation and Research on new policies to advance development of safe and effective cell and gene therapies* www.fda.gov/news-events/press-announcements/ statement-fda-commissioner-scott-gottlieb-md-and-peter-marks-md-phd-director-center-biologics (accessed October 8, 2019).

[41] Journal of Gene Medicine, Gene Therapy Clinical Trials Worldwide Provided by the Journal of Gene Medicine www.abedia.com/wiley/years.php (accessed October 8, 2019).

[42] M. Ruffin and L. Scull, March 28, 2019, *Sector Overview. Gene Therapy for Rare Disorders, ARM*, 2019, https://alliancerm.org/wpcontent/uploads/2019/03/ARM-Gene-Thera py-for-Rare-Disorders_FINAL.pdf (accessed September 29, 2019).

Yet, this cannot account as a decisive indicator since some of the obstacles encountered by gene and cell therapy are intrinsic to how organisms are built, so that largely the same problems accompany their development despite decades of efforts.

In the US, where the majority of the polls have been conducted, the public is benevolent towards gene therapy, especially when intended for severe conditions, while people are ethically alert, wondering about properly informing the subjects, psychological and medical side effects, heritability of modifications, therapeutic applications versus enhancement, resource allocation and potential discrimination.[43] Mentioning risk in terms of potential negative side effects and the danger of eugenics to lay people asked about their opinion on CRISPR-Cas9 technology had an impact, but the respondents generally remained in favor of pursuing with research in the field[44]. Therapeutic editing seems to be supported by the public, both in Europe and the US, while enhancement does not gain the same endorsement[45]; also, interventions in adults are judged as more morally acceptable than interfering with the embryo or the germline.[46]

Still, it seems that public support of genetic modifications in humans decreased over time,[47] a similar decline being noticed in Europe with respect to the use of biotechnologies, even if medical.[48] Could it be related to inappropriate discourse elements eroding people's trust, as "Overhyped promises and exaggerated predictions of the impacts of genetic engineering raised not only hopes, but also suspicion, concerns

[43] J. M. Robillard et al., "Prevailing public perceptions of the ethics of gene therapy," *Human Gene Therapy* 25 (2014): 1-7.

[44] S.M. Weisberg, D. Badgio, and A. Chatterjee, "A CRISPR New World: Attitudes in the Public toward Innovations in Human Genetic Modification," *Frontiers in Public Health* 5 (2017): 1-9.

[45] G. Gaskell et al., "Public views on gene editing and its uses," *Nature Biotechnology* 35 (2017): 1021-1023.

[46] R. J. Blendon, M. T. Gorski, and J. M. Benson, "The Public and the Gene-Editing Revolution," *The New England Journal of Medicine* 374 (2016): 1406-1411. D. A. Scheufele et al., "U.S. attitudes on human genome editing", *Science* 11 (2017): 553-554.

[47] L. Guertin, R McGuire, and A. Torres, "*Public Perception of Human Applications of CRISPR Gene Editing,*" 2018, https://web.wpi.edu/Pubs/E-project/Available/E-project-042618-095556/unrestricted/CRISPR_full_draft.FINAL3.0.pdf (accessed September 29, 2019).

[48] G. Gaskell et al., "Biotechnology and the European public," *Nature Biotechnology* 18 (2000): 935–938.

and anxiety"[49]? Or does the public feel not betrayed but just tired of vain promises? This remains to be established, especially as on the other hand certain data indicate that at least part of the public feels that there is a right to try unregulated therapies.[50] Speaking of this hypothetic right, it should be mentioned that in 2014 The European Court of Human Rights denied a citizen's request to access on behalf of his daughter an unproven stem cell procedure, as it was revealed that the so-called "therapy" was not actually qualifying even for compassionate use, i.e. the use of an unapproved drug when there is no better alternative,[51] which is worrisome, in the context when some countries have adopted Compassionate Use Programmes.[52]

What should be done under the circumstances?

The 2018 Lancet Commission on stem cells and regenerative medicine underlines that "The current gap between public expectations and the realities of translating regenerative technologies threatens regenerative medicine's social 'license to operate'" and asks for stronger regulations, more rigorous publication policies and more responsibility for public dialogue on behalf of researchers and funders.[53]

Scientists should try to communicate better with lay people as social partners entitled to information and decision, as some official documents have already stated.[54] There is a need for developing the required attitude and skills. Some studies revealed that with regard to advanced medical developments (such as stem cells research and the entire topic of regenerative medicine) the larger public has in mind specific concrete aspects pertaining to costs, risks, responsibility and liability, while scientists need to develop their awareness with respect to the same issues

---

[49] H. Gottweis, "Gene therapy and the public: a matter of trust," *Gene Therapy* 9 (2002): 667.

[50] L. Riva et al., "Unproven stem cell therapies: is it my right to try?", *Annali dell'Istituto Superiore di Sanità* 55 (2019): 179-185.

[51] E. Rial-Sebbag and A. Blasimme, "The European Court of Human Rights 'Ruling on Unproven Stem Cell Therapies: A Missed Opportunity?" *Stem Cells and Development* 23 (2014): 39-43.

[52] L. Riva et al., "Unproven stem cell therapies: is it my right to try?", *Annali dell'Istituto Superiore di Sanità* 55 (2019): 179.

[53] G. Cossu et al., "Lancet Commission: Stem cells and regenerative medicine," *Lancet* 391 (2018): 883-910.

[54] European Commission. *Science and Society Action Plan*, 2002. ec.europa.eu/research/swafs/pdf/pub_gender_equality/ss_ap_en.pdf (accessed Sept. 29, 2019). UNESCO. *Recommendation on Science and Scientific Researchers*, 2018. www.unesco.se/wp-content/uploads/2018/11/263618e.pdf (accessed Sept. 29, 2019).

and the social consequences of their work, in general, as well as their interest and availability for science communication, that should also be stimulated by the society.[55]

In any case, exaggeration in assumptions about the potential of rising approaches in medicine should be avoided and the difficulties encountered honestly presented. Not only such an attitude should be ethically imperative, but it might have become a necessary safety measure. There is more than one sort of false promise and some are more dangerous than others.

CONCLUSIONS

In the field of gene and cells therapies, over-enthusiasm had some undesirable consequences. The potential benefits of such approaches were exaggerated and the necessary small steps procedures that incorporate the appropriate safety measures tend on occasions to be eluded. This might indeed threaten regenerative medicine's social "license to operate," as the 2018 Lancet Commission observed.

In the New Age of Science, scientists should be careful about promises and about inducing hopes, as false promises deceive, and false hopes might get to hurt people. Researchers should be permanently aware of the fact that certain developments exert a fascination that can make people vulnerable as it can be easily exploited by rogue scientists and/or other sorts of "confidence artists."

Being open and honest when communicating about the complications that accompany research in modern biotechnologies can maintain the public's appreciation and attract long-term support, while exaggerations and half-truths might exert the opposite effect.

[55] R. Shineha et al., "A Comparative Analysis of Attitudes on Communication toward Stem Cell Research and Regenerative Medicine between the Public and the Scientific Community," *Stem Cells Transl Med* 7 (2018): 251-257. R. Shineha et al., "Science communication in regenerative medicine: Implications for the role of academic society and science policy," *Regen Ther.* 7 (2017): 89-97.

REFERENCES

Barlow, Jane F., and Mo Yang, J. Russell Teagarden. "Are payers ready, willing, and able to provide access to new durable gene therapies?" *Value in Health* 22 (2019): 642–647.

Blendon, Robert J., and Mary T. Gorski, John M. Benson. "The Public and the Gene-Editing Revolution." The New England Journal of Medicine 374 (2016): 1406-1411.

Cossu, Giulio, and Martin Birchall, Tracey Brown, Paolo De Coppi, Emily Culme-Seymour, Sahra Gibbon, Julian Hitchcock et al. "Lancet Commission: Stem cells and regenerative medicine." *Lancet* 391 (2018): 883-910.

Dai, Wei-Jing, and Li-Yao Zhu, Zhong-Yi Yan, Yong Xu, Qi-Long Wang, Xiao-Jie Lu. "CRISPR-Cas9 for in vivo gene therapy: promise and hurdles." Molecular Therapy—Nucleic Acids 5 (2016): 1-4.

Dunbar, Cynthia E., and Katherine A. High, J. Keith Joung, Donald B. Kohn, Keiya Ozawa, Michel Sadelain. "Gene therapy comes of age." *Science* 359 (2018): 1-10.

EMA. ChondroCelect. www.ema.europa.eu/en/medicines/human/EPAR/chondrocelect (accessed September 29, 2019).

EMA. Maci www.ema.europa.eu/en/medicines/human/EPAR/maci (accessed September 29, 2019).

EMA. Product information Alofisel. www.ema.europa.eu/en/documents/product-information/alofisel-epar-product-information_en.pdf (accessed September 29, 2019).

EMA. Product information Glybera. www.ema.europa.eu/en/documents/product-information/glybera-epar-product-information_en.pdf (accessed September 29, 2019).

EMA. Product information Holoclar. www.ema.europa.eu/en/documents/product-information/holoclar-epar-product-information_en.pdf (accessed September 29, 2019)

EMA. Product information Spherox. www.ema.europa.eu/en/documents/product-information/spherox-epar-product-information_en.pdf (accessed September 29, 2019).

EMA. Product information Strimvelis. www.ema.europa.eu/en/documents/product-information/strimvelis-epar-product-information_en.pdf (accessed September 29, 2019).

EMA. Product information Zalmoxis. www.ema.europa.eu/en/documents/product-information/zalmoxis-epar-product-information_en.pdf (accessed September 29, 2019).

EMA. Provenge. www.ema.europa.eu/en/medicines/human/EPAR/provenge (accessed September 29, 2019).

European Commission. Science and Society Action Plan, 2002. https://ec.europa.eu/research/swafs/pdf/pub_gender_equality/ss_ap_en.pdf (accessed September 29, 2019).

FDA. "Approved Cellular and Gene Therapy Products." https://www.fda.gov/vaccines-blood-biologics/cellular-gene-therapy-products/approved-cellular-and-gene-therapy-products (accessed September 29, 2019)

FDA. Investigational New Drug (IND) Application. www.fda.gov/drugs/types-applications/investigational-new-drug-ind-application (accessed October 8, 2019).

FDA. Package Insert and Patient Information PROVENGE®. https://www.fda.gov/media/78511/download (accessed September 29, 2019).

FDA. Package Insert and Patient Information Sheet, LAVIV®. www.fda.gov/media/80838/download (accessed Sept. 29, 2019)

FDA. Package Insert GINTUIT. www.fda.gov/media/83264/download (accessed September 29, 2019).

FDA. Package insert-LUXTURNA. https://www.fda.gov/vaccines-blood-biologics/cellular-gene-therapy-products/luxturna (accessed September 29, 2019).

FDA. Package Insert YESCARTA™. www.fda.gov/media/108377/download (accessed September 29, 2019).

FDA. Package Insert ZOLGENSMA®. www.fda.gov/media/126109/download (accessed September 29, 2019).

FDA. Package Insert, IMLYGIC. www.fda.gov/media/94129/download (accessed September 29, 2019).

FDA. Package Insert, KYMRIAH™. www.fda.gov/media/107296/download (accessed September 29, 2019)

FDA. Package Insert, MACI®. ww.fda.gov/media/101914/download (accessed September 29, 2019).

FDA. Statement from FDA Commissioner Scott Gottlieb, M.D. and Peter Marks, M.D., Ph.D., Director of the Center for Biologics Evaluation and Research on new policies to advance development of safe and effective cell and gene therapies. www.fda.gov/news-events/press-announcements/statement-fda-commissioner-scott-gottlieb-md-and-peter-marks-md-phd-director-center-biologics (accessed October 8, 2019).

Gaskell, George, and Imre Bard, Agnes Allansdottir, Rui Vieira da Cunha, Peter Eduard, Juergen Hampel, Elisabeth Hildt, et al. "Public views on gene editing and its uses." *Nature Biotechnology* 35 (2017): 1021-1023.

Gaskell, George, and Nick Allum, Martin Bauer, John Durant, Agnes Allansdottir, Heinz Bonfadelli, Daniel Boy, et al. "Biotechnology and the European public." *Nature Biotechnology* 18 (2000): 935–938.

Goswami, Reena, and Gayatri Subramanian, Liliya Silayeva, Isabelle Newkirk, Deborah Doctor, Karan Chawla, Saurabh Chattopadhyay, Dhyan Chandra, Nageswararao Chilukuri, and Venkaiah Betapudi. "Gene therapy leaves a vicious cycle." *Frontiers in Oncology* 9 (2019): 1-25.

Gottweis, H. "Gene therapy and the public: a matter of trust." *Gene Therapy* 9 (2002): 667–669.

Grünewald, Julian, and Ronghao Zhou, Sara P. Garcia, Sowmya Iyer, Caleb A. Lareau, Martin J. Aryee, and J. Keith Joung. "Transcriptome-wide off-target RNA editing induced by CRISPR-guided DNA base editors." *Nature* 568 (2019): 433-453.

Guertin, Lauren Robert McGuire, and Alessandra Torres. *Public Perception of Human Applications of CRISPR Gene Editing.* 2018. https://web.wpi.edu/Pubs/E-project/Available/E-project-042618-095556/unrestricted/CRISPR_full_draft.FINAL3.0.pdf (accessed September 29, 2019).

Ihry, Robert J., and Kathleen A. Worringer, Max R. Salick, Elizabeth Frias, Daniel Ho, Kraig Theriault, Sravya Kommineni et al. "p53 inhibits CRISPR–Cas9 engineering in human pluripotent stem cells." *Nature Medicine* 24 (2018): 939–946.

Inoue, Haruhisa, and Naoki Nagata, Hiromi Kurokawa, Shinya Yamanaka. "iPS cells: a game changer for future medicine." *The EMBO Journal* 33 (2014): 409-417.

Kaemmerer, William F. "How will the field of gene therapy survive its success?" *Bioengineering & Translational Medicine* 3 (2018): 166–177.

Kosicki, Michael, and Kärt Tomberg, Allan Bradley. "Repair of double-strand breaks induced by CRISPR–Cas9 leads to large deletions and complex rearrangements." *Nature Biotechnology* 36 (2018): 765-771.

Ma, Guangyu, and Hideaki Shimada, Kenzo Hiroshima, Yuji Tada, Nobuo Suzuki, Masatoshi Tagawa. "Gene medicine for cancer treatment: commercially available medicine and accumulated clinical data in China." *Drug design, Development and Therapy* 2 (2009): 115–122.

Marks, Peter W., and Celia M. Witten, Robert M. Califf. "Clarifying stem-cell therapy's benefits and risks." *The New England Journal of Medicine* 376 (2017): 1007-1009.

McCormack, Mathew P., and Terence H. Rabbitts. "Activation of the T-cell oncogene LMO2 after gene therapy for X-linked severe combined immunodeficiency." *The New England Journal of Medicine* 350 (2004): 913–922.

Raper, Steven E., and Narendra Chirmule, Frank S. Lee, Nelson A. Wivel, Adam Bagg, Guang-ping Gao, James M. Wilson, Mark L. Batshawd. "Fatal systemic inflammatory response syndrome in a ornithine transcarbamylase deficient patient following adenoviral gene transfer." *Molecular Genetics Metabolism* 80 (2003): 148–158.

Rial-Sebbag, Emmanuelle, and Alessandro Blasimme. "The European Court of Human Rights 'Ruling on Unproven Stem Cell Therapies: A Missed Opportunity?" *Stem Cells and Development* 23 (2014): 39-43.

Riva, Luciana, and Laura Campanozzi, Massimiliano Vitali, Giovanna Ricci Vittoradolfo Tambone. "Unproven stem cell therapies: is it my right to try?" *Annali dell'Istituto Superiore di Sanità* 55 (2019): 179-185.

Robillard, Julie M., and Dylan Roskams-Edris, Boris Kuzeljevic, Judy Illes. "Prevailing public perceptions of the ethics of gene therapy." *Human Gene Therapy* 25 (2014): 1-7.

Ruffin, Morrie, and Lyndsey Scull. *Sector Overview. Gene Therapy for Rare Disorders* (2019). https://alliancerm.org/wp-content/uploads/2019/03/ARM-Gene-Therapy-for-Rare-Disorders_FINAL.pdf (accessed September 29, 2019).

Scheufele, Dietram A., and Michael A. Xenos, Emily L. Howell, Kathleen M. Rose, Dominique Brossard, Bruce W. Hardy. "U.S. attitudes on human genome editing." *Science* 11 (2017): 553-554.

Shineha, Ryuma, and Yusuke Inoue, Tsunakuni Ikka, Atsuo Kishimoto, Yoshimi Yashiro. "A Comparative Analysis of Attitudes on Communication toward Stem Cell Research and Regenerative Medicine between the Public and the Scientific Community." *Stem Cells Translational Medicine* 7 (2018): 251-257.

Shineha, Ryuma, and Yusuke Inoue, Tsunakuni Ikka, Atsuo Kishimoto, Yoshimi Yashiro. "Science communication in regenerative medicine: Implications for the role of academic society and science policy." *Regenerative Therapy* 7 (2017): 89-97.

Spoors, John, and Eleanor Croft, Andrew Walker. "Regeneration X: The payer perspective on gene therapy." *British Journal of Healthcare Management* 23 (2017): 158-166.

Stein, Cy A. and Daniela Castanotto. "FDA-Approved oligonucleotide therapies in 2017." *Molecular Therapy* 25 (2017): 1069-1075.

Sung, Yong K. and S. W. Kim. "Recent advances in the development of gene delivery systems." *Biomaterials Research* 12 (2019): 23-28.

Ten Ham, Renske M.T., and Jarno Hoekman, Anke M. Hövels, Andre W. Broekmans, Hubert G.M. Leufkens, Olaf H. Klungel. "Challenges in advanced therapy medicinal product development: A survey among companies in Europe." *Molecular Therapy: Methods & Clinical Development* 11 (2018): 121-130.

The Journal of Gene Medicine. Gene Therapy Clinical Trials Worldwide Provided by the Journal of Gene Medicine www.abedia.com /wiley/years.php (accessed October 8, 2019).

Trounson Alan, and Courtney McDonald. "Stem Cell Therapies in Clinical Trials: Progress and Challenges." *Cell Stem Cell* 17 (2015): 11-22.

UNESCO. Recommendation on Science and Scientific Researchers, 2018. www.unesco.se/wp-content/uploads/2018/11/263618e.pdf (accessed September 29, 2019).

Weisberg, Steven M., and Daniel Badgio, Anjan Chatterjee. "A CRISPR New World: Attitudes in the Public toward Innovations in Human Genetic Modification." *Frontiers in Public Health* 5 (2017): 1- 9.

Zuo, Erwei, Yidi Sun, Wu Wei, Tanglong Yuan, Wenqin Ying, Hao Sun, Liyun Yuan, et al. "Cytosine base editor generates substantial off-target single-nucleotide variants in mouse embryos." *Science* 364 (2019): 289-292.

INFORMED CONSENT. THE CURRENT STANDING

Miroslav Radenković[1]

GENERAL REMARKS

Controlled research clinical trials on human subjects are aimed to test new drugs, involving individual patient recruitment, randomization, and follow-up; they also need to be supported by suitable ethical principles and regulatory background.[2] That way, it is well-accepted that fundamental prerequisites for any ethically grounded research should address the following points[3]:

1. Scientific validity
2. Social value
3. Minimum human risk
4. Benefits should outweigh risks
5. Informed consent
6. Protection of confidentiality and privacy
7. Equitable subject selection with scientific or moral justification for including or excluding subjects from research
8. Protection from harm or exploitation of vulnerable subjects, including children, prisoners, and mentally disabled subjects.

---

[1] Department of Pharmacology, Clinical Pharmacology and Toxicology, Faculty of Medicine, University of Belgrade, Serbia.

[2] Cory E. Goldstein et al., "Ethical Issues in Pragmatic Randomized Controlled Trials: a Review of the Recent Literature Identifies Gaps in Ethical Argumentation," *BMC Medical Ethics* 19(1) (2018), https://doi.org/10.1186/s12910-018-0253-x.

[3] David B. Resnik, "Environmental Health Research Involving Human Subjects: Ethical Issues," *Environmental Health Insights* 2 (2008), https://doi.org/10.4137/ehi.s892.

9. Data integrity and safety monitoring
10. Independent review (Institutional Review Board oversight)

If we are talking about consent in general, it should be applicable to different purposes. That way, consent may be commonly labelled as implied consent, written consent, expressed consent, informed consent, proxy consent, blanket consent, or oral consent.[4] In particular, in research ethics, informed consent represents a course of action of acquiring the consent from the human subject to participate in clinical research after being fully informed of the research aims, study protocol, as well as predictable benefits and potential harms.[5] Throughout this process, a relevant ethics committee is responsible for assuring that the research itself will be conducted in complete accordance with the study plan, adopted methodology, and the highest ethical standards.[6]

Obtaining informed consent from participants in clinical research is essential because it promotes their welfare and ensures their rights.[7] That way, essential ethical principles underlying research on human subjects, specifically autonomy, beneficence, nonmaleficence, and justice, need to be followed.[8] Consequently, we should always be reminded that three essential standards for clinical informed consent must be appreciated: the patient must be competent, adequately informed, and not coerced.[9]

It is therefore evident that informed consent symbolizes a mandatory ethical concept that is extensively implemented in law, as well as in everyday healthcare facility practice. Indeed, clinical informed consent is

---

4 Richa Anand Chandan Dhingra, "Consent in Dental Practice: Patient's Right to Decide," *Journal of Oral Hygiene & Health* 2(2) (2014), https://doi.org/10.4172/2332-0702.1000129.

5 Ana Borovecki et al., "Informed Consent and Ethics Committee Approval in Laboratory Medicine," *Biochemia Medica* 28(3) (2018), doi.org/10.11613/bm.2018.030201.

6 Prashant B. Musmade et al., "Informed Consent: Issues and Challenges," *Journal of Advanced Pharmaceutical Technology & Research* 4(3) (2013): 134, https://doi.org/10.4103/2231-4040.116779.

7 "World Medical Association Declaration of Helsinki," *JAMA* 310(20) (2013): 2191, https://doi.org/10.1001/jama.2013.281053.

8 C. Weijer, "Ethics: Protecting Communities in Biomedical Research," *Science* 289(5482) (2000): 1142-1144, https://doi.org/10.1126/science.289.5482.1142. Raul Artal and Sheldon Rubenfeld, "Ethical Issues in Research," *Best Practice & Research Clinical Obstetrics & Gynaecology* 43 (2017): 107-114, https://doi.org/10.1016/j.bpobgyn.2016. 12.006.

9 C. S. Cocanour, "Informed Consent-It's More than a Signature on a Piece of Paper," *The American Journal of Surgery* 214(6) (2017): 993-997, https://doi.org/10.1016/j.amjsurg. 2017.09.015.

ethically, morally, and legally mandated by the fiduciary responsibilities resulting from the patient-doctor relationship.[10] Finally, as a form of a protective umbrella, specific research ethics committees and institutional review boards assure that participant selection protocols are unbiased, that study participation brings an acceptable ratio of benefits and harms, and that participants provided their informed consent.[11]

INFORM CONSENT – THEN AND NOW

Our present comprehension of informed consent is that it represents the full disclosure of the nature of the research and the human subject's involvement, as well as an adequate understanding of the role of the prospective participant and the participant's voluntary willingness to participate.[12] Unfortunately, from the historical point of view, the conception of informed consent represents the valuable accomplishment of not always peaceful events in both the clinical and research settings in the not so distant past, since in fact, the notion of informed consent is the result of recognized abuses in the clinical and research fields spanning over a century.[13] So ultimately, in order to prevent further exploitations and unhealthy practices in the field of life sciences, the procedure of obtaining informed consent was emphasized to ensure that a human participant's rights were not in danger.[14]

The Nuremberg Code of conducting research on human subjects was introduced after World War II. Namely, following the conclusion of the Nazi Doctors' Trial in 1947, the justices issued the Nuremberg Code with stringent standards to be used in human experimentation, including the

---

10 Timothy J. Paterick et al., "Medical Informed Consent: General Considerations for Physicians," *Mayo Clinic Proceedings* 83(3) (2008): 313-319, https://doi.org/10.4065/83.3.313.

11 Cory E. Goldstein et al., "Ethical Issues in Pragmatic Randomized Controlled Trials."

12 "IRB Consent Form Templates," *IRB Consent Form Templates | Cornell Research Services*, accessed March 26, 2021, https://researchservices.cornell.edu/forms/irb-consent-form-templates.

13 Fida K. Dankar, Marton Gergely, and Samar K. Dankar, "Informed Consent in Biomedical Research," *Computational and Structural Biotechnology Journal* 17 (2019): 463-474, https://doi.org/10.1016/j.csbj.2019.03.010.

14 P. Arun Bhupathi and G.R. Ravi, "Comprehensive Format of Informed Consent in Research and Practice: A Tool to Uphold the Ethical and Moral Standards," *International Journal of Clinical Pediatric Dentistry* 10(1) (2017): 73-81, https://doi.org/10.5005/jp-journals-10005-1411.

premises that human subjects are necessary, that voluntary consent or withdrawal should be obtained, minimal research risks should be present, the research should be terminated if research subjects could be injured or harmed, the research should benefit society, the research must be based on preclinical animal studies, and most importantly, that the research subjects have the right to end participation in the study.[15] Later on, in 1964, the Helsinki Declaration was drafted by the World Medical Association, followed by the US Congress National Research Act of 1974, and the Belmont Report of 1979 underlining three milestone principles: 1. respect for the person; 2. beneficence; and 3. Justice.[16]

Owed to the fast progress in medical sciences, as well as to a variety of new information that has been accumulated, the concept of informed consent has notably changed over time.[17] Although the physician-patient interaction originates from the ethical concept of beneficence, over the nineteenth and twentieth centuries, the law and social changes brought respect for autonomy, too.[18] That way, patients have a legal, ethical and fundamental human right to choose what happens to their bodies, where the consent nowadays represents a form of the patient's agreement for a physician to provide care.[19] So, currently, patient autonomy and the right to choose are the paramount principles, where at the same time patient autonomy counselling has radically progressed, with the ultimate aim that patients should be the key decision-makers, but only if they are fully informed.[20]

Since the practice of obtaining informed consent in the research setting was designed to ensure that the participants' rights were not at risk, research subjects are now better informed and aware of their rights and alternatives, above all, their right of refusal.[21] Unfortunately, present

---

[15] Evelyne Shuster, "Fifty Years Later: The Significance of the Nuremberg Code," *New England Journal of Medicine* 337(20) (1997): 1436-1440, https://doi.org/10.1056/nejm199711133372006.

[16] P. Arun Bhupathi and G.R. Ravi, "Comprehensive Format of Informed Consent in Research and Practice," 73-81.

[17] Fida K. Dankar, Marton Gergely, Samar K. Dankar, "Informed Consent in Biomedical Research," 463-474.

[18] C. S. Cocanour, "Informed Consent—It's More than a Signature on a Piece of Paper."

[19] Akanksha Sood and Janesh Gupta, "Patient Counselling and Consent," *Best Practice & Research Clinical Obstetrics & Gynaecology* 46 (2018): 43-47, https://doi.org/10.1016/j.bpobgyn.2017.10.002.

[20] Akanksha Sood and Janesh Gupta, "Patient Counselling and Consent," 43-47.

[21] Raul Artal and Sheldon Rubenfeld, "Ethical Issues in Research," 107-114.

directives and existing laws cannot predict every reasonable ethical situation or question, particularly given the new advances in science and technology, as well as new and growing medical technology, genetic tests, pharmacological interventions, and other advances which challenge society to retain the highest moral and ethical principles.[22] Yet, the lack of understanding, as well as obstacles in communication, culture, custom, and different other issues, continuously demand an additional concern in obtaining valid informed consent.[23]

THE COMPONENTS OF A VALID CONSENT

When it comes to the components of a valid consent, we always have to be prudent about the challenges related to the decisional capacity, information that will be provided, and the free will concerns.

The capacity

First of all, it must not be presumed that a patient lacks decisional capacity for reaching a decision merely because of their age, disability, behaviour, appearance, medical condition (including mental illness), their apparent inability to communicate, their beliefs, or the fact that they make a choice that a doctor opposes.[24] In other words, a clinician should not doubt a patient's capacity simply because the patient's decision seems unsupported, irrational or based on a series of beliefs that are not in agreement with scientific understandings, peer-reviewed publications or the conventional social norms.[25]

It has to be underlined that the decisional capacity does not primarily coincide with the suitability of the final decision, but more with the pathway through which the participant arrives at that decision, given that patients have diverse capacities to understand and appreciate the information provided, as well to accept and cope with risk and

22 Ibidem.

23 P. Arun Bhupathi and G.R. Ravi, "Comprehensive Format of Informed Consent in Research and Practice," 73-81.

24 Akanksha Sood and Janesh Gupta, "Patient Counselling and Consent," 43-47.

25 S. Henwood, M. A. Wilson, and I. Edwards, "The Role of Competence and Capacity in Relation to Consent for Treatment in Adult Patients," *British Dental Journal* 200(1) (2006): 18-21, https://doi.org/10.1038/sj.bdj.4813118.

uncertainty.[26] Most participants in clinical trials understand the basic elements of informed consent, such as the nature of research, possible benefits of the study, autonomy to withdraw at any time, and the voluntary nature of participation; where at the same time, the understanding of randomization or placebo was shown to be not as accurate.[27]

Children and patients not able to give consent must also be involved in the process, but only to the extent of their capacity to comprehend the study protocol, which indicates that where research involves participants under the age of 18, a consent/permission has to be obtained from parents, and if the child is above 7 years of age then the "child assent" is also required.[28]

When it comes to the frequent questions regarding the competence/capacity relationship, there is also much confusion linked to the appropriate use of these two terms. So, competence is a legal term, and all adults are presumed to be competent unless they are determined by a court to be incompetent, while capacity is task-specific, where decisional capacity is further determined by a clinician.[29] It needs to be emphasized that there may still be concerns in everyday practice, as to a patient's capacity to consent to a proposed treatment, either due to their age or mental capacity.[30] Since capacity is task-specific, it can be divided into eight distinctive areas, detailing the abilities and skills required for each capacity.[31]

In determining whether a patient/participant has the decision-making capacity, the physician must determine if they are capable of these four

---

[26] Nick Nicholas and Mohammed El Sayed, "The Changing Face of Consent: Past and Present," *The Obstetrician & Gynaecologist* 8(1) (2006): 39-44, https://doi.org/10.1576/toag.8.1.039.27207.

[27] Nguyen Thanh Tam et al., "Participants' Understanding of Informed Consent in Clinical Trials over Three Decades: Systematic Review and Meta-Analysis," *Bulletin of the World Health Organization* 93(3) (2015), https://doi.org/10.2471/blt.14.141390.

[28] Prashant B. Musmade et al., "Informed Consent: Issues and Challenges," 134.

[29] Christine S. Cocanour, "Informed Consent—It's More than a Signature on a Piece of Paper," 993-997.

[30] Larner, Elizabeth and Rachel Carter, "The Issue of Consent in Medical Practice," *British Journal of Haematology* 172(2) (2015): 300–304. https://doi.org/10.1111/bjh.13795.

[31] Jennifer Moye and Daniel C. Marson, "Assessment of Decision-Making Capacity in Older Adults: An Emerging Area of Practice and Research," *The Journals of Gerontology: Series B* 62(1) (2007), https://doi.org/10.1093/geronb/62.1.p3.

abilities: understanding, expressing a choice, appreciation, and reasoning[32]:

1. Understanding represents the ability to know the meaning of the information, and it is a key decisional ability.
2. The second ability is whether the person can express a choice.
3. The ability of appreciation is more than just knowing the facts necessary for making a decision. It is about applying those facts to the person's own life.
4. Reasoning refers to the ability to compare options and infer the consequences of the choice.

Information

The process of obtaining informed consent should be characterized by open information exchange.

The patient or study participant should be fully informed about all the risks and probable complications of the administered treatment or investigation offered, where the crucial role of the institutional review board is to ensure that the researcher makes appropriate information available to potential subjects.[33]

Yet, communicating specific technical and other specialized information to those with limited literacy, various social and cultural backgrounds, limited autonomy, and incapacitating diseases may be a difficult mission.[34] As a rule, the information should be provided in lay terms without medical jargon, since it was established that simplified information appeals to patients and is connected with reduced anxiety and increased approval rate of the consent document.[35]

As a useful tool for the understanding of the provided information, different techniques such as the "teach back method," wherein research

[32] Christine S. Cocanour, "Informed Consent—It's More than a Signature on a Piece of Paper," 993-997.

[33] Robert J. Amdur and Elizabeth A. Bankert, *Institutional Review Board: Member Handbook* (Burlington, MA: Jones & Bartlett Learning, 2011).

[34] Rashmi Ashish Kadam, "Informed consent process: A step further towards making it meaningful!" *Perspectives in Clinical Research* 8 (2017): 107-112.

[35] Graham M. Dresden and M. Andrew Levitt, "Modifying a Standard Industry Clinical Trial Consent Form Improves Patient Information Retention as Part of the Informed Consent Process," *Academic Emergency Medicine* 8(3) (2001): 246-252, https://doi.org/10.1111/j.1553-2712.2001.tb01300.x.

participants are asked to say in their own words what has been described, can be successfully employed.[36] Likewise, various questionnaires for testing the level of understanding can be used by introducing simple questions to be answered with[37]:

- yes/no,
- disagree/agree/unsure,
- short answer,
- fill-in the blanks,
- multiple-choice, etc.

The patient's understanding can also be significantly improved by using of printed brochures, particularly those enclosing a mixture of illustrations and narrative content, as well as other various information sheets offering additional information about the clinical trial.[38]

Nowadays, it is more than clear that the use of computers and different multimedia in the consent process may help improve the patient's understanding, given that audio-visual tools have been shown to facilitate the verbal reinforcement of written information, which contributes to the successful comprehension and recall.[39] In other words, the purpose of utilizing multimedia in the consent process is to translate complex study information into a comprehensible and visually appealing audio-video layout.[40]

The use of mobile devices, tablets and the internet for multimedia informed consent delivery is increasingly implemented in everyday routine. Still, whereas social media may boost information exchange, there is a genuine risk of ethical standards deviations, different

[36] Rashmi Ashish Kadam, "Informed consent process," 107-112.

[37] Ibidem.

[38] Robert Michielutte et al., "The Use of Illustrations and Narrative Text Style to Improve Readability of a Health Education Brochure," *Journal of Cancer Education* 7(3) (1992): 251-260, https://doi.org/10.1080/08858199209528176.

[39] Loel Kim et al., "Keeping Users at the Center: Developing a Multimedia Interface for Informed Consent," *Technical Communication Quarterly* 17(3) (2008): 335-357, https://doi.org/10.1080/10572250802100451.

[40] Kathryn Blake et al., "Use of Mobile Devices and the Internet for Multimedia Informed Consent Delivery and Data Entry in a Pediatric Asthma Trial: Study Design and Rationale," *Contemporary Clinical Trials* 42 (2015): 105-118, https://doi.org/10.1016/j.cct.2015.03.012.

misrepresentations, and lack of validity of the information.[41] As an outcome of these concerns, it is strongly recommended that well before market release all novel e-consent tools and consent apps must go through stringent legal and licensing protocols.[42]

Free will

Consent should be given voluntarily. Voluntariness implies that an individual's determination to participate is made without coercion or undue persuasion.[43] Indeed, the person should not be forced in any physical, financial or emotional/mental way to participate in the research.[44] The only person who can consent is the subject themselves.

Consent should always be obtained well ahead, rather than later, as this postponement could sometimes initiate some other delays.[45] At the end of the consent process, an investigator must always ask the participant if they have understood the provided information and if any additional information is necessary to facilitate the decision making process.[46] The physician is obliged to inform the patient of any potentially conflicting interests they may have, including financial or other research-related interests.[47] So, all questions should be answered as fully and honestly as possible, which also means that it is essential to emphasize during the conversation that the subject has the right to refuse to participate in the research or to withdraw from the study at any time after the beginning of the research protocol.[48]

---

41 Celia Taber et al., "Improving the Quality of Informed Consent in Clinical Research with Information Technology," *Studies in health technology and informatics* 231 (2016): 135-142.

42 Roger Collier, "Apple Moves into Medical Research," *Canadian Medical Association Journal* 187(8) (2015): 559-559, https://doi.org/10.1503/cmaj.109-5034.

43 Nguyen Thanh Tam et al., "Participants' Understanding of Informed Consent in Clinical Trials over Three Decades."

44 Ana Borovecki et al., "Informed Consent and Ethics Committee Approval in Laboratory Medicine."

45 O. A. Anderson and I. M. Wearne, "Informed Consent for Elective Surgery--What Is Best Practice?" *Journal of the Royal Society of Medicine* 100(2) (2007): 97-100, https://doi.org/10.1258/jrsm.100.2.97.

46 Akanksha Sood and Janesh Gupta, "Patient Counselling and Consent," 43-47.

47 "Patients' Rights and the Basic Principles of Consent," *Consent Manual*, 43rd Edition (California Hospital Association, 2016), https://www.calhospital.org/sites/main/files/file-attachments/consent2016_webpreview_0.pdf .

48 Ana Borovecki et al., "Informed Consent and Ethics Committee Approval in Laboratory Medicine."

DIFFERENCES BETWEEN INFORMED CONSENT FOR TREATMENT AND RESEARCH PROTOCOLS

Informed consent is not only required for clinical trials, but it is a critical precondition prior to enrolment of each participant in any type of research on human subjects, including diagnostic, therapeutic, interventional, bioequivalence, social, and behavioural studies.[49]

Treatment-related consent

Everyday clinical practice is characterized by the trust relationship between physician and patient, and informed consent obtained for the purpose of treatment is a result of a two-way conversation highlighting the physician-patient therapeutic relationship.[50] Indeed, today's patient-centred medicine underlines the patient's right to actively participate in the healthcare process, where we can clearly recognize a specific discussion between two equal partners (the physician and the patient) about the best possible treatment choices.[51]

In the everyday clinical setting, the patient is always a subject in this process, never an object, where the healthcare provider firstly establishes the patient's condition; then uses lay terms with full consideration of the patient's culture and values, and offers an opportunity for questions related to therapeutic approach and an exchange of important information.[52]

Since the purpose of ethics in the traditional physician-patient relationship is based on doing well, this information-sharing procedure

---

[49] Prashant B. Musmade et al., "Informed Consent: Issues and Challenges," 134.

[50] B.S. Khoury and J.N. Khoury, "Consent: a Practical Guide," *Australian Dental Journal* 60(2) (2015): 138-142, https://doi.org/10.1111/adj.12301. Cory E. Goldstein et al., "Ethical Issues in Pragmatic Randomized Controlled Trials."

[51] Beth A. Ripley et al., "Improving the Informed Consent Conversation: A Standardized Checklist That Is Patient Centered, Quality Driven, and Legally Sound," *Journal of Vascular and Interventional Radiology* 26(11) (2015): 1639-1646, https://doi.org/10.1016/j.jvir.2015.06.007.

[52] Sarah Ford, Theo Schofield, and Tony Hope, "Are Patients' Decision-Making Preferences Being Met?" *Health Expectations* 6(1) (2003): 72-80, https://doi.org/10.1046/j.1369-6513.2003.00211.x. Cathy Charles, Amiram Gafni, and Tim Whelan, "Shared Decision-Making in the Medical Encounter: What Does It Mean? (or It Takes at Least Two to Tango)," *Social Science & Medicine* 44(5) (1997): 681-692, https://doi.org/10.1016/s0277-9536(96)00221-3.

empowers the patient to reach the fully informed decision about accepting or declining the recommended treatment.[53]

However, opposite to the above considerations, there are five recognized exceptions when the informed consent of the patient is not necessary[54]:

- a public health emergency (for example, the use of quarantine)
- a medical emergency (the procedure is needed immediately)
- patient waiver (the patient has requested not to be informed)
- "therapeutic privilege" (a disclosure could seriously harm, rather than benefit the patient)
- in case the patient is incompetent (a need for determination of a surrogate)

Research-related consent

The term *research* is traditionally linked to a class of activities designed to further contribute to general knowledge.[55] During this process, the investigator's responsibility is to obtain informed consent from potential human subjects, and moreover, it has to be stressed that the research community significantly depends on the integrity of the investigator to complete the consent process successfully.[56]

The relationship between the researcher and the participants is guided by a specific series of rules, all with the rationale of confirming a scientific hypothesis[57]:

- the selection of suitable subjects
- appropriate treatment of subjects,
- careful application of research methodology,

---

[53] B.S. Khoury and J.N. Khoury, "Consent: a Practical Guide," 138-142.

[54] Christine S. Cocanour, "Informed Consent—It's More than a Signature on a Piece of Paper," 993-997.

[55] "International Ethical Guidelines for Health-Related Research Involving Humans • COUNCIL FOR INTERNATIONAL ORGANIZATIONS OF MEDICAL SCIENCES," *COUNCIL FOR INTERNATIONAL ORGANIZATIONS OF MEDICAL SCIENCES*, October 12, 2020, https://cioms.ch/shop/product/international-ethicalguidelines-for-health-related-research-involving-humans.

[56] Robert J. Amdur and Elizabeth A. Bankert, *Institutional Review Board: Member Handbook.*

[57] Ana Borovecki et al., "Informed Consent and Ethics Committee Approval in Laboratory Medicine."

- acquisition of high-quality data.

In this scenario, high ethical standards are equated with the high quality of science, which means that informed consent in research implies that participants give the researcher permission to actually treat them as an object of the research.[58] So one must be very careful of the so-called therapeutic misconception, which means that study participants may mistakenly believe that research activities are meant to directly benefit them and their personal health.[59]

The main models of informed consent with their related characteristics are presented below[60]:

Consent model	Characteristics
Standard informed consent	• Written disclosure of all information. • Consent obtained in writing.
Targeted consent model	• Verbally disclosed information. • Obtained written consent.
Integrated consent model	• Verbally disclosed information. • No written consent required. • Physician documents consent in electronic health record.
Streamlined consent	• No consent is sought.

Today, as opposed to some novel interpretations of informed consent, most researchers support without any doubt the notion that irrespective of the research risk levels, written record of the patient's agreement to participate in the proposed research still remains exceptionally important.[61] This way, we are effectively protecting a patient's decision-making role, and at the same time we are respecting

58 Ibidem.

59 Cory E. Goldstein et al., "Ethical Issues in Pragmatic Randomized Controlled Trials."

60 Ibidem.

61 James R. Anderson, Toby L. Schonfeld, "Informed consent for comparative effectiveness trials," *The New England journal of medicine* 370(20) (2014), https://doi.org/10.1056/NEJMc1403310#SA1.

the person's dignity.[62] After all, it is assumed that an informed consent obtained in a conventional mode generates satisfaction, reduces inadequate expectations and provides documented proof of good clinical practice.[63]

Finally, there are some circumstances where informed consent can be waived; however, waivers of consent require justification, including the following examples:

1. the research must be socially important;
2. requiring informed consent would make the study infeasible;
3. participation poses only minimal risk.[64]

DISCLOSURE OF INFORMATION

It is crucial to disclose all information about a study to the participants.[65] This usually includes the following items: 1. the nature of the research, 2. the purpose of the research, 3. risks and anticipated benefits, 4. any available alternative procedures, and 5. the opportunity for the subject to ask any questions.[66]

Specifically, the current regulations require that prospective research participants receive the following information[67]:

---

[62] Ahmed Elsayyad, "Informed Consent for Comparative Effectiveness Trials," *New England Journal of Medicine* 370(20) (2014): 1958-1960, https://doi.org/10.1056/nejmc 1403310#SA3.

[63] J.M. Mellado, "Autonomía, Consentimiento y Responsabilidad. Parte II. El Consentimiento Informado En La Medicina Asistencial y En La Jurisprudencia," *Radiología* 58(6) (2016): 427-434, https://doi.org/10.1016/j.rx.2016.06.010.

[64] "International Ethical Guidelines for Health-Related Research Involving Humans • COUNCIL FOR INTERNATIONAL ORGANIZATIONS OF MEDICAL SCIENCES."

[65] Office for Human Research Protections, "Federal Policy for the Protection of Human Subjects ('Common Rule," *HHS.gov* (US Department of Health and Human Services, March 18, 2016), https://www.hhs.gov/ohrp/regulations-and-policy/regulations/ common-rule/index.html.

[66] Office for Human Research Protections, "The Belmont Report," *HHS.gov* (US Department of Health and Human Services, March 15, 2016), www.hhs.gov/ohrp/ regulations-and-policy/belmont-report. Charles Weijer et al., "The Ottawa Statement on the Ethical Design and Conduct of Cluster Randomized Trials," *PLoS Medicine* 9(11) (2012), https://doi.org/10.1371/journal.pmed.1001346.

[67] Office for Human Research Protections (OHRP), "45 CFR 46," HHS.gov (US Department of Health and Human Services, March 10, 2021), www.hhs.gov/ ohrp/humansubjects/guidance/45cfr46.html.

- a statement that the study involves research;
- the explanation of its purpose and procedures;
- the description of the reasonably foreseeable risks or discomforts;
- the description of the possible benefits to the participant or society, where monetary compensation is not a benefit;
- the description of alternative treatments that might be advantageous to the subject;
- the expected duration of the study
- the explanation of the steps taken to protect the confidentiality of private health information, including a description of who may have access to research records;
- for research posing more than minimal risk, an explanation as to whether compensation will be available for research-related injury;
- explaining whom to contact about information on the research or the participant's rights;
- a statement that research participation is voluntary and study refusal (or withdrawal) will not result in the loss of benefits to which the participant has a right.

When appropriate, the following information must also be provided[68]:

- Unforeseeable risks: A statement that the particular treatment or procedure may involve risks to the subject (or to the embryo or fetus, if the subject is or may become pregnant) that are currently unforeseeable.
- Termination of participation by the investigator: Anticipated circumstances under which the subject's participation may be terminated by the investigator without regard to the subject's consent.
- Additional costs: Any additional costs to the subject that may result from participation in the research.
- Consequences of discontinuing research participation: The consequences of a subject's decision to withdraw from the research; and procedures for orderly termination of participation by the subject.

[68] Robert J. Amdur and Elizabeth A. Bankert, *Institutional Review Board: Member Handbook.*

- Notification of significant new findings: A statement that significant new findings developed during the course of the research that may relate to the subject's willingness to continue participation will be provided to the subject.
- An approximate number of subjects: The approximate number of subjects involved in the study.

FINAL REMARKS

Informed consent should always be considered as much more than a patient's signature on a legal document.

Informed consent must be viewed as a continuous dynamic process rather than an isolated event during the clinical study.

It is essential to adopt innovative multimedia communication strategies to further enhance comprehension of clinical trial information among participants.

Improving the usefulness and efficiency of the informed consent process is a critical healthcare and research policy issue.

Shifting the conversation from the national regulations to the internationally accepted ethical principles should be a continual process.

Developing international ethical standards for responsible conduct of clinical research trials requires the participation of all.

ACKNOWLEDGEMENTS

The Fogarty International Centre at the US National Institutes of Health (NIH) and the Ministry of Education and Science of the Republic of Serbia supported this article through grants 2R25 TW008171-06A1, and 175073, respectively.

REFERENCES

Goldstein, Cory E., and Charles Weijer, Jamie C. Brehaut, Dean A. Fergusson, Jeremy M. Grimshaw, Austin R. Horn, Monica Taljaard. "Ethical Issues in Pragmatic Randomized Controlled Trials: a Review of the Recent Literature Identifies Gaps in Ethical Argumentation." *BMC Medical Ethics* 19(1) (2018). https://doi.org/10.1186/s12910-018-0253-x.

Resnik, David B. "Environmental Health Research Involving Human Subjects: Ethical Issues." *Environmental Health Insights* 2 (2008). https://doi.org/10.4137/ehi.s892.

Chandan Dhingra, Richa Anand. "Consent in Dental Practice: Patient's Right to Decide." *Journal of Oral Hygiene & Health* 2(2) (2014). https://doi.org/10.4172/2332-0702.1000129.

Borovecki, Ana, and Ana Mlinaric, Martina Horvat, Vesna Supak Smolcic. "Informed Consent and Ethics Committee Approval in Laboratory Medicine." *Biochemia Medica* 28(3) (2018). https://doi.org/10.11613/bm.2018.030201.

Musmade, Prashant B, and LokeshP Nijhawan, N Udupa, KL Bairy, KM Bhat, ManthanD Janodia, BS Muddukrishna. "Informed Consent: Issues and Challenges." *Journal of Advanced Pharmaceutical Technology & Research* 4(3) (2013): 134. https://doi.org/10.4103/2231-4040.11 6779.

"World Medical Association Declaration of Helsinki." *JAMA* 310(20) (2013), https://doi.org/10.1001/jama.2013.281053.

Weijer, C. "ETHICS: Protecting Communities in Biomedical Research." *Science* 289 (2000): 1142–44. https://doi.org/10.1126/science.289. 5482.1142.

Artal, Raul, and Sheldon Rubenfeld. "Ethical Issues in Research." *Best Practice & Research Clinical Obstetrics & Gynaecology* 43 (2017): 107–14. https://doi.org/10.1016/j.bpobgyn.2016.12.006.

Cocanour, Christine S. "Informed Consent – It's More than a Signature on a Piece of Paper." *The American Journal of Surgery* 214(6) (2017): 993–97. https://doi.org/10.1016/j.amjsurg.2017.09.015.

Paterick, Timothy J., Geoff V. Carson, Marjorie C. Allen, and Timothy E. Paterick. "Medical Informed Consent: General Considerations for Physicians." *Mayo Clinic Proceedings* 83(3) (2008): 313–319. https://doi.org/10.4065/83.3.313.

"IRB Consent Form Templates." *IRB Consent Form Templates | Cornell Research Services.* Accessed March 26, 2021. https://researchservices. cornell.edu/forms/irb-consent-form-templates. Accessed date: 8 April 2020.

Dankar, Fida K., and Marton Gergely, Samar K. Dankar. "Informed Consent in Biomedical Research." *Computational and Structural Biotechnology Journal* 17 (2019): 463–74. https://doi.org/10.1016/j.c sbj.2019.03.010.

Bhupathi, P. Arun, and G.R. Ravi. "Comprehensive Format of Informed Consent in Research and Practice: A Tool to Uphold the Ethical and Moral Standards." *International Journal of Clinical Pediatric Dentistry* 10(1) (2017): 73–81. https://doi.org/10.5005/jp-journals-10005-1411.

Shuster, Evelyne. "Fifty Years Later: The Significance of the Nuremberg Code." *New England Journal of Medicine* 337(20) (1997): 1436–40. https://doi.org/10.1056/nejm199711133372006.

Sood, Akanksha, and Janesh Gupta. "Patient Counselling and Consent." *Best Practice & Research Clinical Obstetrics & Gynaecology* 46 (2018): 43–47. https://doi.org/10.1016/j.bpobgyn.2017.10.002.

Henwood, S., and M. A. Wilson, I. Edwards. "The Role of Competence and Capacity in Relation to Consent for Treatment in Adult Patients." *British Dental Journal* 200(1) (2006): 18–21. https://doi.org/10.1038/sj.bdj.4813118.

Nicholas, Nick, and Mohammed El Sayed. "The Changing Face of Consent: Past and Present." *The Obstetrician & Gynaecologist* 8(1) (2006): 39–44. https://doi.org/10.1576/toag.8.1.039.27207.

Tam, Nguyen Thanh, Nguyen Tien Huy, Le Thi Thoa, Nguyen Phuoc Long, Nguyen Thi Trang, Kenji Hirayama, and Juntra Karbwang. "Participants' Understanding of Informed Consent in Clinical Trials over Three Decades: Systematic Review and Meta-Analysis." *Bulletin of the World Health Organization* 93(3) (2015). https://doi.org/10.2471/blt.14.141390.

Larner, Elizabeth, and Rachel Carter. "The Issue of Consent in Medical Practice." British Journal of Haematology 172(2) (2015): 300–304. https://doi.org/10.1111/bjh.13795.

Moye, Jennifer, and Daniel C. Marson. "Assessment of Decision-Making Capacity in Older Adults: An Emerging Area of Practice and Research." *The Journals of Gerontology: Series B* 62(1) (2007). https://doi.org/10.1093/geronb/62.1.p3.

Amdur, Robert J., and Elizabeth A. Bankert. *Institutional Review Board: Member Handbook*. Burlington, MA: Jones & Bartlett Learning, 2011.

Kadam, Rashmi Ashish. "Informed consent process: A step further towards making it meaningful!" *Perspectives in Clinical Research* 8 (2017): 107-112.

Dresden, Graham M., and M. Andrew Levitt. "Modifying a Standard Industry Clinical Trial Consent Form Improves Patient Information

Retention as Part of the Informed Consent Process." *Academic Emergency Medicine* 8(3) (2001): 246–52. https://doi.org/10.1111/j.1553-2712.2001.tb01300.x.

Michielutte, Robert, and Judy Bahnson, Mark Dignan, Elissa Schroeder. "The Use of Illustrations and Narrative Text Style to Improve Readability of a Health Education Brochure." *Journal of Cancer Education* 7(3) (1992): 251–60. https://doi.org/10.1080/08858199209528176.

Kim, Loel, and Amanda J. Young, Robert A. Neimeyer, Justin N. Baker, Raymond C. Barfield. "Keeping Users at the Center: Developing a Multimedia Interface for Informed Consent." *Technical Communication Quarterly* 17(3) (2008): 335–57. https://doi.org/10.1080/10572250802100451.

Blake, Kathryn, Janet T. Holbrook, Holly Antal, David Shade, H. Timothy Bunnell, Suzanne M. McCahan, Robert A. Wise, Chris Pennington, Paul Garfinkel, and Tim Wysocki. "Use of Mobile Devices and the Internet for Multimedia Informed Consent Delivery and Data Entry in a Pediatric Asthma Trial: Study Design and Rationale." *Contemporary Clinical Trials* 42 (2015): 105–18. https://doi.org/10.1016/j.cct.2015.03.012.

Taber C, Warren J, Day K. "Improving the Quality of Informed Consent in Clinical Research with Information Technology." *Studies in health technology and informatics* 231 (2016): 135-142.

Collier, Roger. "Apple Moves into Medical Research." *Canadian Medical Association Journal* 187(8) (2015): 559–59. https://doi.org/10.1503/cmaj.109-5034.

Anderson, O. A, and I M. Wearne. "Informed Consent for Elective Surgery--What Is Best Practice?" *Journal of the Royal Society of Medicine* 100(2) (2007): 97–100. https://doi.org/10.1258/jrsm.100.2.97.

"Patients' Rights and the Basic Principles of Consent." *Consent Manual 2016*. 43rd Edition. California Hospital Association, 2016. https://www.calhospital.org/sites/main/files/file-attachments/consent2016_webpreview_0.pdf. Accessed on 8 April 2020.

Khoury, B.S., and J.N. Khoury. "Consent: a Practical Guide." *Australian Dental Journal* 60(2) (2015): 138–42. https://doi.org/10.1111/adj.12301.

Ripley, Beth A., and David Tiffany, Lisa S. Lehmann, Stuart G. Silverman. "Improving the Informed Consent Conversation: A Standardized Checklist That Is Patient Centered, Quality Driven, and

Legally Sound." *Journal of Vascular and Interventional Radiology* 26(11) (2015): 1639–46. https://doi.org/10.1016/j.jvir.2015.06.007.

Ford, Sarah, and Theo Schofield, Tony Hope. "Are Patients' Decision-Making Preferences Being Met?" *Health Expectations* 6(1) (2003): 72–80. https://doi.org/10.1046/j.1369-6513.2003.00211.x.

Charles, Cathy, and Amiram Gafni, Tim Whelan. "Shared Decision-Making in the Medical Encounter: What Does It Mean? (or It Takes at Least Two to Tango)." *Social Science & Medicine* 44(5) (1997): 681–92. https://doi.org/10.1016/s0277-9536(96)00221-3.

"International Ethical Guidelines for Health-Related Research Involving Humans • Council For International Organizations Of Medical Sciences." *Council For International Organizations Of Medical Sciences*, October 12, 2020. https://cioms.ch/shop/product/international-ethicalguidelines-for-health-related-research-involving-humans. Accessed 02 Apr 2020.

Anderson, James R., and Schonfeld, Toby L. "Informed consent for comparative effectiveness trials." *The New England Journal of Medicine*, 370(20) (2014). https://doi.org/10.1056/NEJMc1403310#SA1.

Elsayyad, Ahmed. "Informed Consent for Comparative Effectiveness Trials." *New England Journal of Medicine* 370(20) (2014): 1958–60. https://doi.org/10.1056/nejmc1403310#SA3.

Mellado, J.M. "Autonomía, Consentimiento y Responsabilidad. Parte II. El Consentimiento Informado En La Medicina Asistencial y En La Jurisprudencia." *Radiología* 58(6) (2016): 427–34. https://doi.org/10.1016/j.rx.2016.06.010.

Office for Human Research Protections. "Federal Policy for the Protection of Human Subjects ('Common Rule." HHS.gov. US Department of Health and Human Services, March 18, 2016), https://www.hhs.gov/ohrp/regulations-and-policy/regulations/common-rule/index.html. Accessed date: April 08, 2020.

Office for Human Research Protections. "The Belmont Report." HHS.gov. US Department of Health and Human Services, March 15, 2016. https://www.hhs.gov/ohrp/regulations-and-policy/belmont-report. Accessed date: April 08, 2020.

Weijer, Charles, and Jeremy M. Grimshaw, Martin P. Eccles, Andrew D. McRae, Angela White, Jamie C. Brehaut, Monica Taljaard. "The Ottawa Statement on the Ethical Design and Conduct of Cluster

Randomized Trials." *PLoS Medicine* 9(11) (2012). https://doi.org/10.1371/journal.pmed.1001346.

Office for Human Research Protections (OHRP). "45 CFR 46." HHS.gov. US Department of Health and Human Services, March 10, 2021. www.hhs.gov/ohrp/humansubjects/guidance/45cfr46.html.

INFORMED CONSENT AND MEDICAL DECISION MAKING: ETHICAL CHALLENGES

Raluca Dumache,[1] Alexandra Enache[2]

INTRODUCTION

In medical practice, the relationship between the doctor and their patient represents the basis of successful clinical medicine, thus improving the patients' health.[3] An essential factor that needs to be considered when talking about the relationship between a doctor and their patients is confidentiality, based on all the principles of medical ethics: autonomy, beneficence, non-maleficence and justice.

THE IMPORTANCE OF INFORMED CONSENT AND ITS ETHICAL IMPLICATIONS

The patient's informed consent is compulsory in any medical act that implies risks for the patient. The patient authorizes the medical act before it is performed and supposes that the patient has received all the information referring to his/her medical status.[4]

---

[1] "Victor Babes" University of Medicine and Pharmacy, Department of Legal Medicine, Bioethics, Deontology and Medical Law; and "Victor Babes" University of Medicine and Pharmacy, Centre of Ethics in Human Genetic Identification, Timisoara Romania.

[2] "Victor Babes" University of Medicine and Pharmacy, Department of Legal Medicine, Bioethics, Deontology and Medical Law; and Victor Babes University of Medicine and Pharmacy, Centre of Ethics in Human Genetic Identification, Timisoara, Romania.

[3] S. H. Kaplan, "Characteristics of physicians with participatory decision-making styles," *Ann Intern Med* 124 (1996): 497–04.

[4] M.A. Stewart, "Effective physician patient communication and health outcomes: a review," *CMAJ* 152 (1995):1423–33.

In a medical environment, consent represents a person's willing agreement – a person who possesses sufficient mental capacity to be able to take a decision referring to a medical procedure or to a treatment suggested by a specialized medical member[5]

Informed consent is the process by which the healthcare provider informs a competent patient about the appropriate treatment, allowing the patient to decide voluntarily to accept or refuse a medical procedure for the treatment.[6] When there are two or more treatment options, autonomy and the competent patient has the absolute right to be informed and to decide the treatment to which he/she would be subjected to, being fully informed about all the risks and benefits.[7]

Consent can be:

- Expressed, verbally or in written form.
- Implicit, the presumption of the agreement.

INFORMING THE PATIENT

Every patient is a citizen with full rights, thus beyond the disease state that determines the patient to ask for specialized help, these rights must be respected and should not diminish his/her illness, and on the contrary, they can be emphasized for the sake of the patient`s protection.[8]

Certain aspects that raise a dilemma, both from an ethical and judicial point of view, can harden the doctor's decision referring to the patient:

- Informing the patient about a potentially serious/lethal disease vs. not informing the patient – therapeutic privilege.
- The doctor has the right not to inform the patient if they believe that the patient's health or life is in danger when they find out the diagnosis.

[5] T.L. Beauchamp, "Informed consent: Its history, meaning, and present challenges," *Camb Q Health Ethics* 20 (2001): 515–523.

[6] C. Grady, "Enduring and emerging challenges of informed consent," *N Engl J Med* 372 (2015):855–862.

[7] Paul S. Appelbaum, "Assessment of Patients' Competence to Consent to Treatment," *N Engl J Med* 357 (2007): 1834–1840.

[8] D. Rothman, *Strangers at the bedside: a history of how law and ethics transformed medical decision making* (New York: Basic Books, 1991).

- The doctor has the right not to inform the patient who does not express the wish to find out the relevant information regarding their state of health or does not want to take decisions on their own.
- The doctor has the right not to inform the patient who lacks autonomy, understanding capacity, or has decisional competence.

We cannot talk about the therapeutic privilege in the following situations:

- The patient is independent and wants to know the truth about their state of illness.
- The patient suffers from a contagious disease that implies an obligatory declaration.
- The patient needs other medical or surgical interventions for which they must express their informed consent.[9]

PATIENT AUTONOMY

Based on self-determination, each and every person having and being in their full state of mind can take decisions and administrate their own life and interests.

From a bioethical point of view, a person is independent when they are psychological competent,

- To understand their situation
- To be voluntary
- To have decisional capacity.

The autonomy of the person has three meanings:

- The bioethical principle based on the free decision in the medical practice.
- The philosophical principle of the existence of a rational being.
- The state of a person to present their psychic capacity and to be in the legal state of exercising their civil rights at the same time, that is to say, to be psychic competent.

[9] A.E. Buchanan, D.W. Brock, *Deciding for Others: The Ethics of Surrogate Decision Making* (Cambridge: Cambridge University Press, 1990), 187-134.

MEDICAL CONFIDENTIALITY

Confidentiality is not an absolute concept and there are situations in which the doctor has to infringe this right of the patient. The Tarasoff Case (1976) was one of these, which raised numerous questions referring to the idea of confidentiality and its limits.[10] Prosenjit Poddar, while treated by a psychiatrist, confessed to him of his intention to hurt Tatiana Tarasoff, and even though the doctor asked the authorities to arrest him after showing them the seriousness of the situation through the diagnosis of paranoid schizophrenia, Poddar was not imprisoned as he was not considered dangerous. Poddar eventually murdered Tatiana Tarasoff, and the victim`s family sued the doctor invoking that they were not informed about Poddar`s plans that put Tatiana's life in danger.

The confidentiality of the medical data is the patient's right and at the same time, the doctor's obligation.

When the patient addresses the doctor and presents them with the reasons their specialized help was sought for, these reasons are to be private for each and every patient, and they have the following implications:

- moral – they are confidential and are deontological and ethical protected by confidentiality.
  The doctor will keep the professional secret and will act according to the legal right of each person to the respect of his private life referring to the information about his health.[11]
- legal – they are a professional secret protected by criminal and organic laws.
  All the information referring to the patient`s health, the results of the investigations, the diagnosis, the prognosis, the treatment, the personal data are confidential[12]

Professional secrets in the deontology of the medical practice is called confidentiality.

This confidentiality is the base of the relation of trust between the doctor and their patient, and it also contributes to the quality of the medical act due to the possibility to create a therapeutic alliance.

[10] R. Weinstock, G. Vari, G.B. Leong et al., "Back to the past in California: A temporary retreat to a Tarasoff duty to warn," *J Am Acad Psychiatry Law* 34 (2006): 523–528.
[11] *The Deontological Code of the Romanian Doctors' College 2017*, Chapter III, art. 17.
[12] *The Law of the Patient's Rights* 2012, Chapter 4, art. 21.

The patient's confidence helps the doctor obtain information about the symptoms and the history of the disease so that the doctor can organize their knowledge and thinking in order to offer the best treatment.

It is a legal, moral, and deontological obligation for the doctor to respect the professional secrets in medicine. On the other hand, keeping the confidentiality strengthens and emphasizes the relation between the doctor and the patient, and this situation contributes to the stability of the therapeutic alliance and the patient's compliance, thus contributing to the quality of the medical act. It must be mentioned that regarding the patient's medical confidentiality, we cannot discuss subjectivism or the judicial system creating laws for that, including the Criminal Code; thus, the infringement of the confidentiality becomes not only a simple deontological violation, but also represents a criminal act. We must also add that there is no transparency in the confidentiality of the medical data except when the law especially requires it.

There are cases foreseen by the law that requires breaking the professional secret. When the judicial institutions ask the doctor to break the professional secret and the confidentiality of the patient's medical data – in the case the legal requirements for this action are met – the doctor will ask for a written demand and a closed door meeting.

In this article, we present the results of a 10-question survey which was presented to people during March 2018–March 2019. The study consisted of 200 participants, age 18–65 years old, having different jobs.

Statistical analysis:

For the statistical part, we used the χ^2 test. p-values of 0.05 and 0.001 were used to distinguish between different levels of statistical significance. In all cases, all questions of the survey had p-value < 0.001, with a statistical significance.

Demographic characteristics of the participants:

Participants were recruited online through Facebook. Participants included were 18 years and older. 146 participants (73%) were female and 54 participants (27%) were male, as presented in Fig. 1.

The participants were divided into 6 groups depending on their age, as follows: 18–25 years: 113 participants (56.5%); 26–35 years: 34 (17%); 36–45 years: 29 participants (14.5%); 46–55 years: 14 participants (7%); 56–65 years: 7 participants (3.5%); 3 persons (15%) were older than 65 years. This information is presented in Fig. 2.

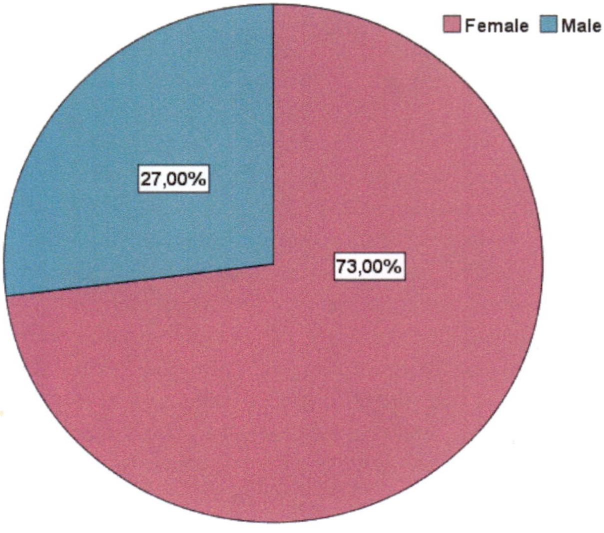

Fig. 1. Gender distribution of the participants

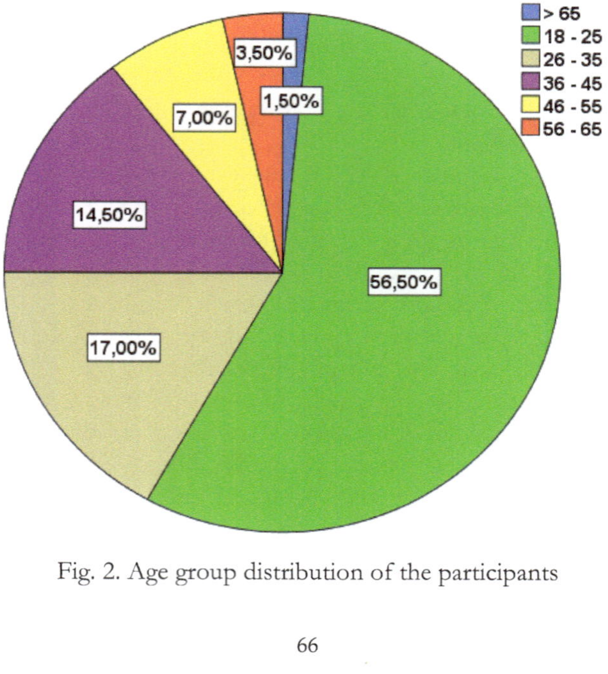

Fig. 2. Age group distribution of the participants

Table 1. Survey questions

Question number	**Question**	Answer: **Yes**	Answer: **No**	Answer: **Do not know**
Q1	Have you completed/signed an informed consent?	143 (71.5%)	44 (22%)	13 (6.5%)
Q2	Has the medical professional explained why you need informed consent?	129 (64.5%)	57 (28.5%)	14 (7%)
Q3	Do you believe that the explanations received from the medical staff were clear?	141 (70.5%)	41 (20.5%)	18 (9%)
Q4	Did the medical professional present the risks of medical/surgical treatment?	142 (71%)	43 (21.5%)	15 (7.5%)
Q5	Do you think you would need more explanation for the risks?	79 (39.5%)	98 (49%)	23 (11.5%)
Q6	Did the medical professional present you with alternative treatment options?	121 (60.5%)	66 (33%)	13 (6.5%)
Q7	Do you think your medical information is kept confidential?	108 (54%)	23 (11.5%)	69 (34.5%)
Q8	Are you concerned about the confidentiality of your medical data?	125 (62.5%)	69 (34.5%)	6 (3%)
Q9	According to your knowledge, does a person have the right to see their medical health record?	159 (79.5%)	14 (7%)	27 (13.5%)

Q10	Have you ever requested a consultation or a copy of your personal medical health record?	45 (22.5%)	150 (75%)	5 (2.5%)

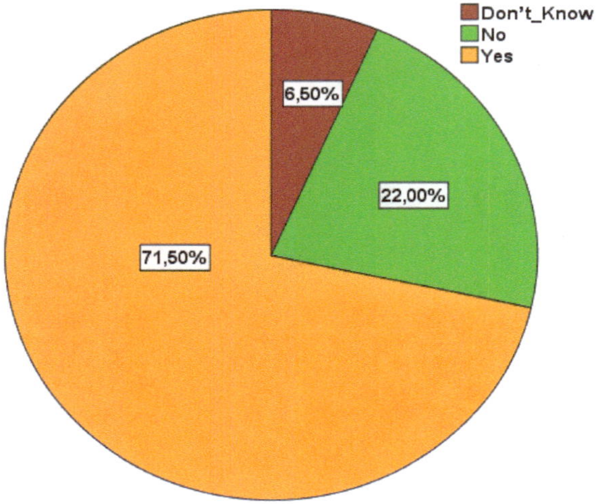

Fig.3. *Q1* Have you completed/signed an informed consent?

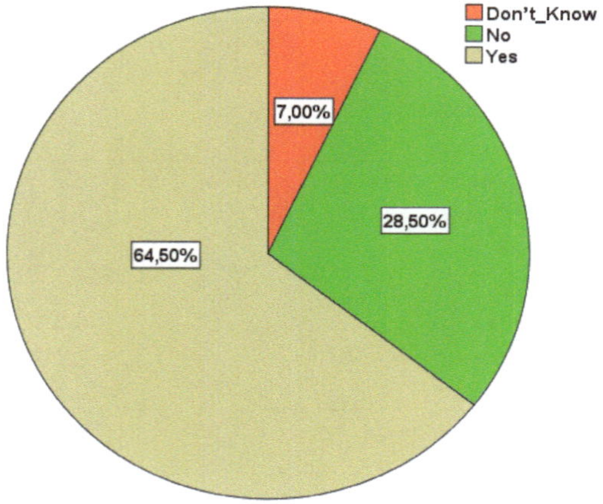

Fig.4. *Q2* Has the medical professional explained why you need informed consent?

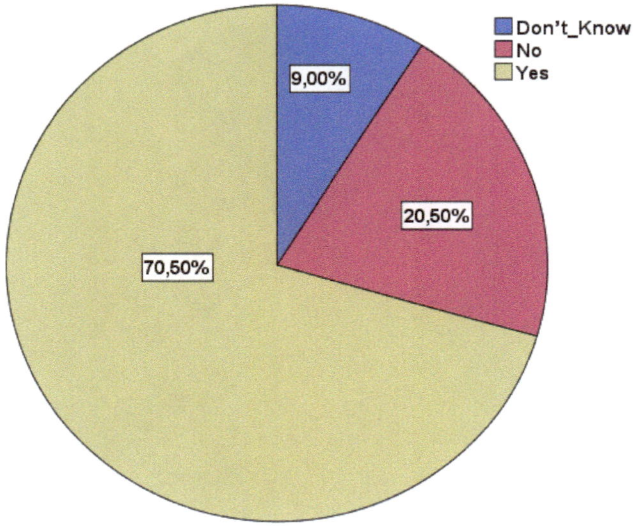

Fig. 5. *Q3* Do you believe that the explanations received from the medical staff were clear?

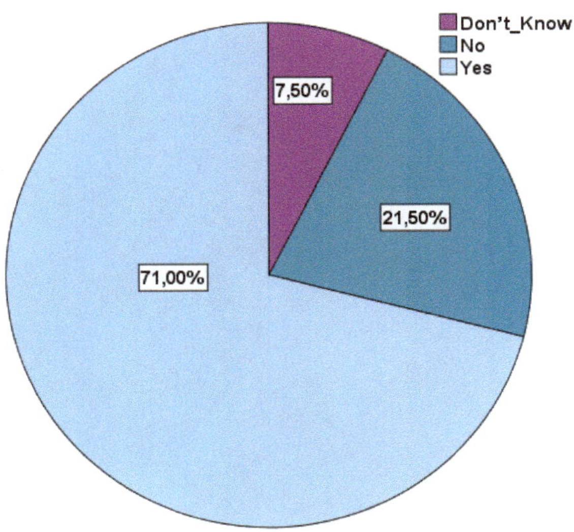

Fig. 6. *Q4* Did the medical professional present the risks of medical/surgical treatment?

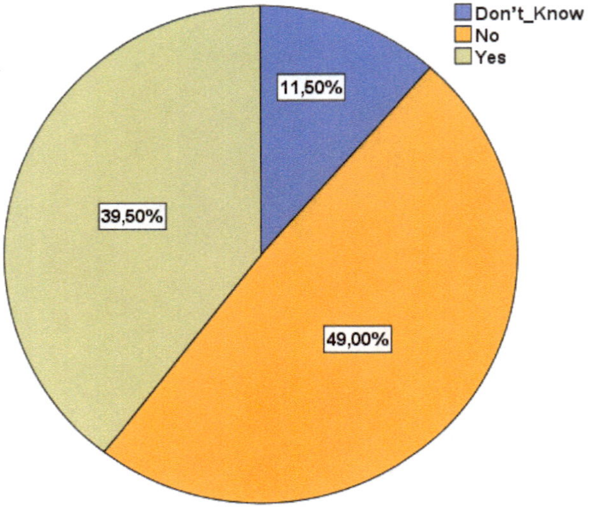

Fig. 7. *Q5* Do you think you would need more explanation for the risks?

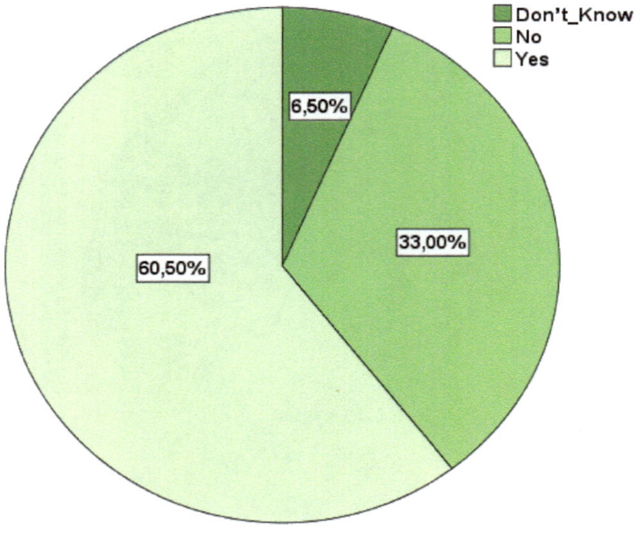

Fig. 8. *Q6* Did the medical professional present you with alternative treatment options?

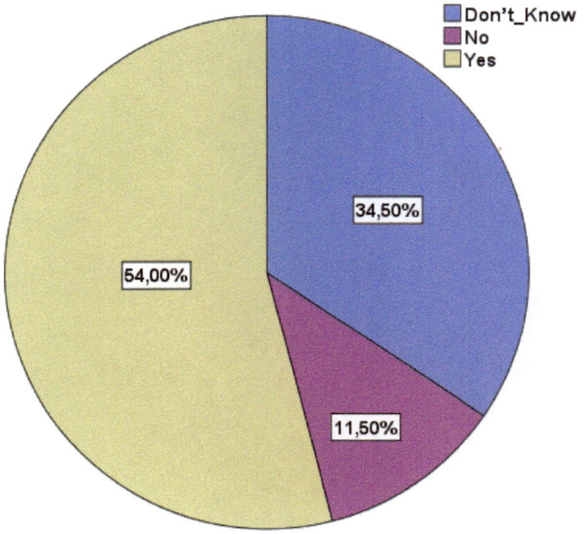

Fig. 9. *Q7* Do you think your medical information is kept confidential?

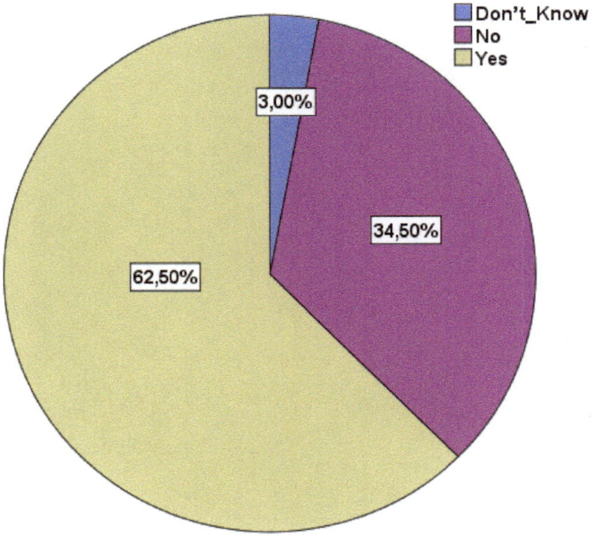

Fig. 10. *Q8* Are you concerned about the confidentiality of your medical data?

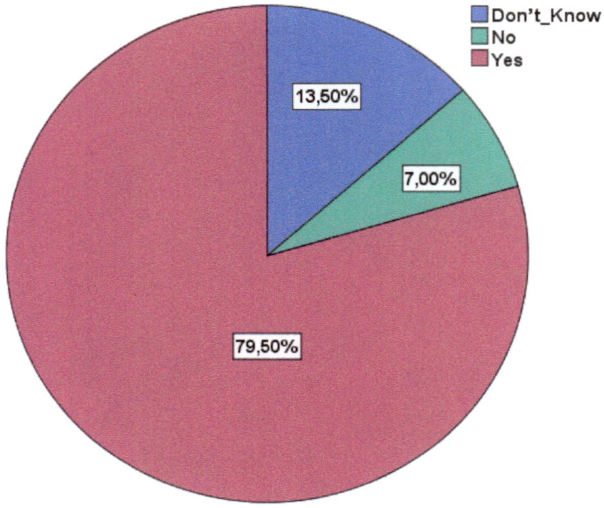

Fig. 11. *Q9* According to your knowledge, does a person have the right to see their medical health record?

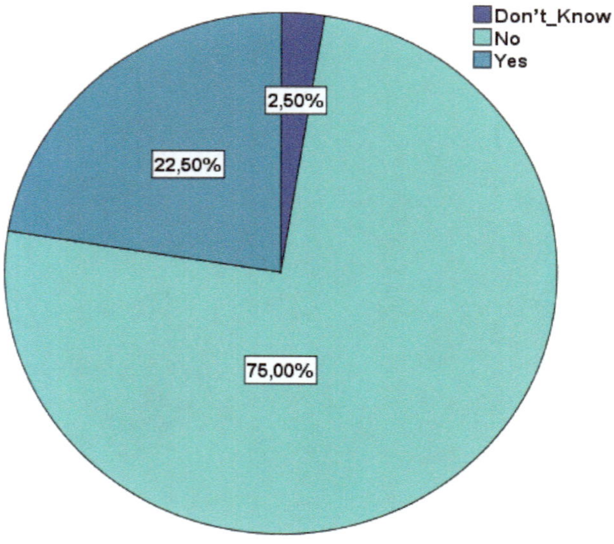

Fig. 12. Q10 Have you ever requested a consultation or a copy of your personal medical health record?

The majority of the participants confirmed that they had signed an informed consent (71.5% – 143 persons), but they also mentioned the fact that the physician offered them the complete information regarding the informed consent (64.5% – 129 persons) (Fig. 3). Regarding the feedback related to the explications that resulted from the discussions between the physician, more than half of the questioned persons considered the explanations to be very clear (70.5% – 141 persons), detailed (54% – 108 persons) or enough (63.3% – 126 persons) (Fig. 4). 33.5% of the questioned persons felt that that explanations could be more detailed, 26.6% were not sufficient to aid them, and 20.5% responded that the explanations were not sufficient or easy to understand (Fig. 5).

Regarding the risks that can occur during a medical or surgical procedure or treatment, they have to be informed about these by the physician. Only 71% (142 persons) from the questioned persons confirmed that they had been informed about these risks. A percentage of 21.5% (43 persons) denied this, while 13 persons (7.5%) did not know what to answer (Fig. 6). Almost half of the persons who completed the questionnaire – 98 persons (49%) – said that regarding the risks, they did not need to be informed, but a percentage of 39.5% (79 persons) said that the physician did not offer clear information (Fig. 7).

Regarding the treatment, other therapeutic options, or the responsibility of the decision maker, the majority of the questioned persons (121 persons – 60.5%) said that they received information about other treatments, while 124 persons (62.0%) felt involved in the decision making related to the proposed treatment (Fig. 8). The confidentiality of the medical information became an important subject for those who need medical care, and in this case, more than half of the questioned persons – 125 participants (62.5%) – are worried about this; meanwhile about 108 persons (54%) believe their medical information is kept safely (Fig. 9, Fig. 10).

When the participants were asked about the right of the patient to see their medical records, 159 persons (79.5%) agreed with this, while 14 persons (7%) did not agree with this. Also, 27 persons (13.5%) did not know what to answer.

When participants were asked if they asked for their medical records, 75% – 150 participants – did not ask for their medical records, while 45 persons (22.5%) asked for their medical records, and only 5 persons (2.5%) did not know what to answer.

CONCLUSION

The first priority in a committed doctor-patient relationship is considering the well-being of the patient, involving the patient in all the medical decisions that concern their health, and respecting the patient's decisions, thus respecting their autonomy.

From the responses of the survey, we can conclude that the participants were concerned about the informed consent and the explanations received from the medical staff about their health and well-being. Also, participants are very involved in medical decision-making when it comes to the recommendations about the medical treatment. Last but not at least, people are concerned about the confidentiality of their medical data.

To our knowledge, this is the first study in Romania related to the attitudes of the general population toward informed consent, the confidentiality of medical data, and medical decision making.

DECLARATION OF INTEREST

The authors declare there is no conflict of interest related to this article.

ACKNOWLEDGEMENTS

We gratefully thank the "Victor Babes" University of Medicine and Pharmacy Timisoara, Romania, for the financial support in conducting the project titled: *Perfectible medical education through quality improvement of teaching in line with academic ethics and integrity*. CNFIS-FDI_2019_0396

REFERENCES

Appelbaum, Paul S. "Assessment of Patients' Competence to Consent to Treatment." *N Engl J Med* 357 (2007): 1834–1840.

Beauchamp, T.L. "Informed consent: Its history, meaning, and present challenges." *Camb Q Health Ethics* 20(4) (2001): 515–523.

Buchanan, A.E., and D.W. Brock. *Deciding for Others: The Ethics of Surrogate Decision Making*. Cambridge: Cambridge University Press, 1990.

Grady, C. "Enduring and emerging challenges of informed consent." *N Engl J Med.* 372(9) (2015): 855–862.

Kaplan, S.H. "Characteristics of physicians with participatory decision-making styles." *Ann Intern Med.* 124(4) (1996): 497-504.

Legare, F., and S. Turcotte, D. Stacey, S. Ratte, J. Kryworuchko, I.D. Graham. "Patients' perceptions of sharing in decisions: a systematic review of interventions to enhance shared decision making in routine clinical practice." *Patient* 5(1): 2012.

Rothman, D. *Strangers at the bedside: a history of how law and ethics transformed medical decision making.* New York: Basic Books, 1991.

Stewart, M.A. "Effective physician patient communication and health outcomes: a review." *CMAJ* 152 (1995): 1423–33.

The Deontological Code of the Romanian Doctors' College. Chapter III, art. 17.

The Law of the Patient's Rights 2012. Chapter 4, art. 21.

Weinstock, R., and G. Vari, G.B. Leong, et al. "Back to the past in California: A temporary retreat to a Tarasoff duty to warn." *J Am Acad Psychiatry Law* 34(4) (2006): 523–528.

PHARMACOTHERAPY OF RARE DISEASES IN SERBIA: BIOETHICAL CHALLENGES AND HOW TO OVERCOME THEM

Branislava Medić Brkić, Miroslav Radenković, Katarina Savić Vujović, Nevena Divac, Sonja Vučković, Radan Stojanović, Dragana Srebro, Miloš Basailović and Milica Prostran[1]

BACKGROUND

A rare disease (RD) is defined as a disease or disorder that affects 1 in 1,500 people, according to the U.S. National Institutes of Health -NIH. In comparison, the European Union defines rare diseases as those affecting less than 1 in 2,000 people or 5 in 10,000 citizens.[2]

While any rare disease will affect a small number of people, when taken together, all rare diseases (there are over 7,000 today) affect more than 25 million Americans and over 30 million European Union citizens. Although there are no official data, it is estimated that in Serbia half a million people suffer from some of these conditions.[3]

---

[1] Department of Pharmacology, Clinical Pharmacology and Toxicology Faculty of Medicine, University of Belgrade, Serbia.

[2] E. Adams, J. Deutsche, E. Okoroh, S. Owens-Mcalister, S. Majumdar, M. Ullman, M-L. Damiano, M. Recht, L. Casto, K. Droze, "An inventory of healthy weight practices in federally funded haemophilia treatment centres in the United States," *Haemophilia: the official journal of the World Federation of Hemophilia* 20(5) (2014): 639-43. K. Ahmad, A. Daliri, J. Mamikhani, "Cost-effectiveness of prophylaxis against on-demand treatment in boys with severe haemophilia in Iran," *Value Health* 12(3) (2009): A133.

[3] Rare Diseases Europe (EURORDIS): *About Rare Diseases*, www.eurordis.rs/aboutrare-diseases (accessed: June 15, 2020). Serbian National Organization for Rare Diseases (NORBS): *About Rare Diseases* (Serbian), www.norbs.rs/norbso-retkim-bolestima.php (accessed: June 15, 2020).

We must note that Serbia has made certain progress concerning this topic in the last decade. The National Organization for Rare Diseases in Serbia (NORBS) was formed as a non-governmental, non-profit organization in 2010 with the main focus to improve the quality of life of patients who suffer from RDs in our country. This organization is the main ideological driver of all actions in the country that aim to advance the health, social and bioethical aspects of these patients. Besides this, progress has been made in the laws concerning these disorders. The law on Social protection in 2011 provided four different kinds of support for these persons: financial social assistance, allowance and extended allowance for help and care of another person and one-time financial assistance. Additionally, in 2014, the first legislative framework for patients with rare diseases was made. These amendments suggested the need for the creation of centres for rare diseases, promoted these patients as a special category with a claim to use free health insurance and formed an additional budget in the Ministry of Health in order to invest in orphan drugs.[4]

Still, there are many obstacles and unresolved issues. For example, until today, there is no proper coding system for most RDs in Serbia, accurate and well-timed diagnostics in many cases is still missing, there is no sufficient number of registered orphan drugs, the procedures for their procurement are complicated, etc.

Because of all of the above, education on rare diseases contributes not only to more accurate and timely diagnoses and treatment of particular patients, but also improves the general aspects of this problem such as planning a national strategy, funding, foundation of expert centres etc. Therefore, the role of medical, but also bioethical education is essential in providing a good foundation for present and future doctors involved in this issue.

[4] *Amandments to the Law on Health Care. Official Gazette Republic of Serbia*, no. 57/2011. (Belgrade: Official Gazzete of the Republic of Serbia, 2011). www.parlament.gov.rs/narodna-skupstina-.871.html (accessed: June 3, 2020). M. Joldic, J. Todorovic, Z. Terzic-Supic, "The needs of patients with rare disease in Serbia: Why do we need National Strategy for rare disease?" *Health & Social Care in the Community* (2019): 1-10. Branislava Medić Brkić, Bojan Stopić, Katarina Savić Vujović, Nevena Divac, Sonja Vučković, Radan Stojanović, Dragana Srebro, Miloš Basailović, Milica Prostran, "Pharmacotherapy of Rare Diseases in Serbia: The Current State of Art," in "*Rare Diseases*," ed. Zhan He Wu (IntechOpen), DOI: 10.5772/intechopen.91262. www.intechopen.com/books.

BARRIERS IN THE PHARMACOTHERAPY OF RARE DISEASES

Patients affected by rare diseases, in general, face many problems. They are often misdiagnosed or the diagnosis is delayed due to several problems: lack of awareness among physicians, lack of expertise, unavailability and/or high costs of diagnostic tests and drugs for rare diseases (orphan drugs). According to the National Organization of Patients with Rare Diseases in Serbia (NORBS), many diagnostic procedures have to be conducted abroad and the process comprises many difficulties: high costs, travel expenses or transportation of biological material etc. Although national legislation ensures the availability of drugs for those diseases, pharmacotherapy is faced with a lot of problems. Licensed drugs are not always or not readily available and provision of unlicensed medicines is very difficult due to legislative obstacles.[5]

The reason for this situation is the fact that pharmaceutical companies are not motivated to invest in the development and distribution of low-consumption drugs on the market. They need to invest a significant financial resource in orphan drugs but often struggle to recover costs through sales. Besides this, there are other obstacles regarding drugs for rare diseases. There are no Orphan Drug Regulations and cross-sector sponsorship.[6]

Additionally, clinical trials that examine drugs for rare diseases also face many difficulties and limitations. Performing clinical studies properly, in this case, is difficult because the number of potential participants is low, there are numerous challenges regarding uniform diagnostic criteria, lack of reliable patient registries, use of placebo group

[5] B. Medić, N. Divac, B. Stopić, K. Savić Vujović, A. Glišić, N. Cerovac, M. Prostran, "The attitudes of medical students towards rare diseases: A cross-sectional study," *Vojnosanitetski Pregled* 73 (2015): 94.

[6] M. E. Haffner, "Adopting orphan drugs – two dozen years of treating rare diseases," *N Engl J Med* 354 (5) (2006): 445–7. R. Fiorentino, G. Liu, A. R. Pariser, A. E. Mulberg, "Cross-sector sponsorship of research in eosinophilic esophagitis: a collaborative model for rational drug development in rare diseases," *J Allergy Clin Immunol* 130 (3) (2012): 613-6. M. Dunoyer, "Accelerating access to treatments for rare diseases," *Nat Rev Drug Discov* 10 (7) (2011): 475–6.

(which may not be ethical because of the severity of some rare diseases), technical problems to design this kind of clinical trials etc.[7]

In addition to clinicians, researchers are also not particularly interested in dealing with and performing research on rare diseases and orphan drugs. There is a logical explanation for this. It is difficult to obtain relevant and reliable data on rare diseases because of data heterogeneity, outcomes and measurement techniques may differ from one country to another, even for the same diseases. The second problem represents the dissemination of the results of some RD examinations. Journal editors may hesitate to publish studies with a narrow audience and small impact.[8]

BIOETHICAL CHALLENGES REGARDING PHARMACOTHERAPY OF RARE DISEASES

There is a lot of bioethical controversy and open questions regarding rare diseases and their pharmacotherapy. Research on rare diseases should satisfy the ethical conditions that apply to human subjects generally.

Patients suffering from RDs could be recognized as vulnerable subjects. It is well known that most of these diseases represent chronic, degenerative, and life-threatening conditions with an unknown or genetic origin. Besides this, more than half of all patients are children.[9]

Some RDs could have an overwhelming influence on all aspects of the patients' lives and their careers. Also, the development and availability of orphan drugs could threaten basic bioethical principles.

---

[7] R. J. Lilford, J. G. Thornton, D. Braunholtz, "Clinical trials and rare diseases: a way out of a conundrum," *BMJ (Clinical research ed).* 311(7020) (1995): 1621–5. M. Luisetti, I. Campo, R. Scabini, M. Zorzetto, Z. Kadija, F. Mariani, I. Ferrarotti, "The problems of clinical trials and registries in rare diseases," *Respir Med* 104 (Suppl 1) (2010): S42–4. S. Groft, P. de la Paz, "Rare diseases – avoiding misperceptions and establishing realities: the need for reliable epidemiological data," *Adv Exp Med Biol* 686 (2010): 3–14.

[8] M. Pai, C. H. T. Yeun, E. A. Akl, "Strategies for eliciting and synthesizing evidence for guidelines in rare diseases," *BMC Med Res Methodol* 19(1) (2019): 67.

[9] Z. Todorović, M. Prostran, B. Medić, M. Vučinić, "Bioehics and pharmacology," in *Bioethics and Pharmacology: Ethics in Preclinical and Clinical Drug Development*, eds. Z. Todorović, M. Prostran, K. Turza, K. Kerala (Transworld Research Network, 2012), 7-13. M. Prostran, Z. Todorović, R. Stojanović, T. Potpara, Z. Nešić, J. Lazić, B. Medić, "Bioethics in clinical trials: Vulnerable subjects," in *Bioethics and Pharmacology: Ethics in Preclinical and Clinical Drug Development*, eds. Z. Todorović, M. Prostran, K. Turza, K. Kerala (Transworld Research Network, 2012), 87-100.

However, participation in clinical trials and research on rare diseases can affect each of the four basic bioethical principles. We have previously discussed four commonly accepted principles of medical ethics regarding the issue of rare diseases, excerpted from Beauchamp and Childress. They include the: (1) principle of respect for autonomy, (2) principle of non-maleficence, (3) principle of beneficence, and (4) principle of justice.[10] At this point, we will point out some of the most important and challenging facts regarding this topic.

• Autonomy and informed consent

The lack of alternatives can significantly affect patients' free will to participate in rare disease research. This is especially important when patients need to sign an informed consent in order to participate in a clinical trial for orphan drugs.

In addition, we must emphasize here that a large percentage of these diseases are of genetic origin and that the largest number of patients are children, who cannot make decisions independently.

• Beneficence and maleficence

Clinical studies and research on rare diseases can challenge this bioethical principle in several ways. We have already discussed the dangers of including a placebo control group of patients in this type of study. However, there are other limitations such as specific interventions and methods of administering these substances, insufficiently reliable data on such therapies, insufficiently trained experts and adequate medical equipment. Clinical trials for rare and ultra-rare diseases typically involve a limited number of research sites recruiting from a small pool of patients dispersed over a large geographical area. In this situation, individuals and their families are required to travel great distances, often at high cost personally and financially.[11]

• Justice

The bioethical principle in this kind of research is often considered in relation to participant selection and ensuring that the risks and benefits

---

[10] T. Beauchamp, J. Childress, *Principles of Biomedical Ethics*, 7th ed. (New York: Oxford University Press, 2013): 10-17.

[11] L. Gelinas, B. Crawford, A. Kelman, B. E. Bierer, "Relocation of study participants for rare and ultra-rare disease trials: Ethics and operations," *Contemp Clin Trials* 84 (2019): 105812

of research participation are distributed equitably among individuals and social groups.[12]

However, in many cases, this is not possible because there are no standards and guidelines for many RDs or their differences from one county to another. In these situations, an offer of relocation and participation in some trials may be the best option available, but the question is whether this opportunity will be equally accessible to all. The answer is probably no.

THE IMPACT OF BIOMEDICAL EDUCATION ON ATTITUDES REGARDING RARE DISEASES

We considered it crucial to examine different aspects of knowledge and attitudes regarding rare disease issues in Serbia. In order to do this, we performed several surveys among medical students, medical doctors and pharmacists, as well as among a group of patients suffering from rare neurological diseases.[13]

Generally, our respondents were highly interested in expressing their attitudes regarding the issue of rare diseases and their treatment in our country and we had a high percentage of provided answers.

A survey conducted among third- and sixth-year medical students showed that final-year students did not make significant progress regarding knowledge and attitudes on rare diseases and their pharmacotherapy in comparison to their younger fellows. Nevertheless, their answers indicated that they understand the problem and the importance of dealing with this issue in our country.

In the second part of our examination, our respondents from the physicians and pharmacists voluntarily participated in our survey. Although these were health professionals who meet patients with rare

---

[12] National Commission for the Protection of Human Subjects of Biomedical and Behavior.al Research The Belmont Report. https://www.hhs.gov/ohrp/regulations-and-policy/belmont-report/read-the-belmont-report/index.html (accessed: June 1, 2020).

[13] B. Medic, B. Stopic, S. Peric, V. Rakocevic Stojanovic, D. Lavrnic, K. Savic Vujovic, S. Vuckovic, N. Divac, R. Stojanovic, D. Srebro, M. Prostran, "How to Live with Rare Diseases in Serbia: The Knowledge and Attitudes of Patients Regarding Medical, Societal and Bioethical Aspects of this Issue," *International Journal of Humanities Social Sciences and Education* 3(2) (2016): 24-30. B. Medić, Z. Todorović, K. Savić Vujović, R. Stojanović, M. Prostran, "Oboleli od retkih bolesti kao vulnerabilni ispitanici u kliničkim studijama," *JAHR* 8(4)(2013): 715-724.

diseases, most of them believed they are not sufficiently educated on this issue. However, they recognized the key problems of such patients in our community: lack of scientific knowledge, complicated procedures for the procurement of drugs, the lack of a sufficient number of registered medicines for these diseases etc.

Finally, in the third part of our work, we performed a cross-sectional study among outpatients and inpatients (Neurology Clinic, Department for Neuromuscular Disorders, Clinical Center of Serbia) who suffered from rare neurological diseases. Results showed that only half of the participants knew that they suffered from a rare disease and in general they have shown very poor knowledge on this subject in our country. Besides this, only 16.6% of examined patients were part of some rare diseases patients' organizations and just a quarter of our respondents (26.6%) agreed to participate in clinical trials concerning orphan drugs. The improvement of the pharmacotherapy of rare diseases in Serbia, medical professionals and patients can be achieved by enhancement of general and scientific knowledge, well-timed and appropriate diagnostics, simplified procedures for drug procurement, registration of more orphan drugs, etc.

CONCLUSION

Our research has shown that it is necessary to improve knowledge and attitudes among all participants of the health system in the country (present and future medical doctors and pharmacists, patients and their families, regulatory bodies and general population).

We firmly believe that this is the right time to start a comprehensive and constructive debate in society on this topic, both from a medical and legal perspective. Each of these participants should make a contribution in order to improve the position of this group of patients.

The first step is education that should relate not only to medical and legal issues, but also to the ethical issues of patients with rare diseases. As professors at the Medical Faculty of the largest university in Serbia, we believe that the beginning of this education should have started among students of medicine and pharmacy, but also among students of law and philosophy. This is the critical point for conducting a public discussion in order to significantly improve the position of rare diseases patients' in our community in future.

ACKNOWLEDGEMENTS

This work was supported by the Ministry of Education, Science and Technological Development of Serbia (Grant 175023).

REFERENCES

Adams, E., and Deutsche J., Okoroh E., Owens-Mcalister S., Majumdar S., Ullman M., Damiano M.L., Recht M., Casto L., Droze K. "An inventory of healthy weight practices in federally funded haemophilia treatment centres in the United States." *Haemophilia: the official journal of the World Federation of Hemophilia* 20(5) (2014): 639-43.

Ahmad K., and Daliri A., Mamikhani J. "Cost-effectiveness of prophylaxis against on-demand treatment in boys with severe haemophilia in Iran." *Value Health* 12(3) (2009): A133.

Amendments to the Law on Health Care. Official Gazette Republic of Serbia 2011, no. 57/2011. Belgrade: Official Gazzete of the Republic of Serbia. www.parlament.gov.rs/narodna-skupstina-.871.html. Accessed: June 3, 2020.

Beauchamp T., and Childress J. *Principles of Biomedical Ethics*, 7th ed. New York: Oxford University Press, 2013.

Dunoyer, M. "Accelerating access to treatments for rare diseases." *Nat Rev Drug Discov* 10 (7) (2011): 475-6.

Fiorentino, R., and Liu G., Pariser A.R., Mulberg A.E. "Cross-sector sponsorship of research in eosinophilic esophagitis: a collaborative model for rational drug development in rare diseases." *J Allergy Clin Immunol* 130 (3) (2012): 613-6.

Gelinas L., and Crawford B., Kelman A., Bierer B.E. "Relocation of study participants for rare and ultra-rare disease trials: Ethics and operations." *Contemp Clin Trials* 84 (2019): 105812. doi: 10.1016/j.cct.2019.105812.

Groft, S., and de la Paz P. "Rare diseases - avoiding misperceptions and establishing realities: the need for reliable epidemiological data." *Adv Exp Med Biol* 686 (2010): 3-14.

Haffner, M.E. "Adopting orphan drugs--two dozen years of treating rare diseases." *N Engl J Med* 354 (5) (2006): 445-7.

Joldic, M., and Todorovic J., Terzic-Supic Z. "The needs of patients with rare disease in Serbia: Why do we need National Strategy for rare disease?" *Health & Social Care in the Community* (2019): 1-10.

Lilford, R.J., and Thornton J.G., Braunholtz D. "Clinical trials and rare diseases: a way out of a conundrum." *BMJ (Clinical research ed).* 311(7020) (1995): 1621-5.

Luisetti, M., and Campo I., Scabini R., Zorzetto M., Kadija Z., Mariani F., Ferrarotti I. "The problems of clinical trials and registries in rare diseases." *Respir Med* 104 (Suppl 1) (2010): S42-4.

Medić Brkić, B., and Stopić B., Savić Vujović K., Divac N., Vučković S., Stojanović R., Srebro D., Basailović M., Prostran M. "Pharmacotherapy of Rare Diseases in Serbia: The Current State of Art", in *Rare Diseases,* ed. Zhan He Wu. IntechOpen. DOI: 10.5772/intechopen.91262. https://www.intechopen.com/books.

Medić, B., and Divac N., Stopić B., Savić Vujović K., Glišić A., Cerovac N., Prostran M. "The attitudes of medical students towards rare diseases: A cross-sectional study." *Vojnosanitetski Pregled* 73 (2015).

Medic B., and Stopic B., Peric S., Rakocevic Stojanovic V., Lavrnic D., Savic Vujovic K., Vuckovic S., Divac N., Stojanovic R., Srebro D., Prostran M. "How to Live with Rare Diseases in Serbia: The Knowledge and Attitudes of Patients Regarding Medical, Societal and Bioethical Aspects of this Issue." *International Journal of Humanities Social Sciences and Education* 3(2) (2016): 24-30.

Medić B., and Todorović Z., Savić Vujović K., Stojanović R., Prostran M. "Oboleli od retkih bolesti kao vulnerabilni ispitanici u kliničkim studijama." *JAHR* 8(4) (2013): 715-724.

National Commission for the Protection of Human Subjects of Biomedical and Behavioral Research The Belmont Report. https://www.hhs.gov/ohrp/regulations-and-policy/belmont-report/read-the-belmont-report/index.html. Accessed: June 1, 2020.

Pai, M., and Yeung C.H.T., Akl E.A. "Strategies for eliciting and synthesizing evidence for guidelines in rare diseases." *BMC Med Res Methodol* 19(1) (2019).

Prostran M., and Todorović Z., Stojanović R., Potpara T., Nešić Z., Lazić J., Medić B. "Bioethics in clinical trials: Vulnerable subjects." In *Bioethics and Pharmacology: Ethics in Preclinical and Clinical Drug Development*, ed. Todorović Z., Prostran M., Turza K., 87-100. Transworld Research Network, 2012.

Rare Diseases Europe (EURORDIS): *About Rare Diseases.* www.eurordis.rs/aboutrare-diseases. Accessed: June 15, 2020.

Serbian National Organization for Rare Diseases (NORBS). *About Rare Diseases* (Serbian). www.norbs.rs/norbso-retkim-bolestima.php. Accessed: June 15, 2020.

Todorović, Z., and Prostran M., Medić B., Vučinić M. "Bioehics and pharmacology." In *Bioethics and Pharmacology: Ethics in Preclinical and Clinical Drug Development*, ed. Todorović Z., Prostran M., Turza K., 7-13. Transworld Research Network, 2012.

ETHICAL ISSUES REGARDING CHRONIC PAIN FROM THE PERSPECTIVE OF A CLINICAL PHARMACOLOGIST

Katarina Savić Vujović,[1] Miroslav Radenković,[1] Branislava Medić,[1] Slađana Mihajlović,[2] Sonja Vučković,[1] Nevena Divac,[1] Radan Stojanović,[1] Dragana Srebro,[1] Miloš Basailović,[1] Milica Prostran[1]

INTRODUCTION

Pain is the most common symptom of a disease. It is defined by the International Association for the Study of Pain (IASP) as "an unpleasant sensory and emotional experience associated with actual or potential tissue damage, or described in terms of such damage."[3] Qualified specialists in the field of pain are responsible for adequate and successful pain treatment as an ethical imperative and obligation. Acute and chronic pain are different clinical entities. Acute pain is pain that appears suddenly, can be severe, but lasts a relatively short period of time. Acute pain is usually induced by inflammation, trauma or injury.[4] The aim of therapy for acute pain is to treat the underlying cause of pain and to

[1] Department of pharmacology, clinical pharmacology and toxicology, Faculty of Medicine, University of Belgrade, Belgrade, Serbia.
[2] Ob/Gyn Hospital, University Clinical Centar "Dr Dragisa Misovic Dedinje", Belgrade Serbia.
[3] Rolf-Detlef Treede, "The International Association for the Study of Pain definition of pain: as valid in 2018 as in 1979, but in need of regularly updated footnotes," *Pain Rep* 3(2) (2018): e643.
[4] KP Grichnik, FM Ferrante, "The difference between acute and chronic pain," *Mt Sinai J Med* 58(3) (1991): 217-20.

interrupt the nociceptive signals. Improperly treated, acute pain increases the body's neuroendocrine and immune responses to stress and can produce early and late adverse systemic effects and complications, with retardation of healing, increased morbidity, mortality, prolonged hospitalization and the transition of acute to chronic pain.[5]

Chronic pain persists over a longer time than acute pain and is usually resistant to most medical treatments.[6] Chronic pain is defined as persistent or recurrent pain lasting longer than three months.[7] It often causes severe problems for patients. Chronic pain may be considered a disease state. Chronic pain occurs in about 20% of the adult population, and more often in women and the older population.[8] Pain often lasts because of inadequate treatment, and many patients suffer for months and years. This condition leads to depression, anxiety, professional disability and deconditioning.[9] Insomnia is very common in people with chronic pain. It is a consequence of the mechanism of central sensitization, characterized by a state of neuroinflammation and commonly associated with stress or anxiety. Usually, these patients do not respond effectively to the therapy.[10]

The treatment of chronic pain is very difficult. There are many discrepancies between the diagnosis and the treatment of chronic pain with regard to ethical issues that must be addressed. The ethical issues associated with chronic pain treatment are numerous. Ethical principles such as autonomy, integrity, beneficence, equity and nonmaleficence should be considered when treatment of chronic pain is considered.

---

[5] K. Kirsh, A.K. Vice, S. Passik, "History of Opioids and Opiophobia," in *Pain and chemical dependency*, ed. H. Smith, S. Passik (New York: Oxford University Press, 2008), 3-8.

[6] Sonja Vucković, Milica Prostran, Milovan Ivanović, Zoran Todorović, Radan Stojanović, Zorica Nesić, Ivana Matić, Slobodan Milovanović, "Opioid analgesics," *Vojnosanit Pregled* 61(4) (2004): 13-21.

[7] Daniel Z Buchman, Anita Ho, Judy Illes, "You present like a drug addict: patient and clinician perspectives on trust and trustworthiness in chronic pain management," *Pain Med* 17(8) (2016): 1394–1406.

[8] O. van Hecke, N. Torrance, B.H. Smith, "Chronic pain epidemiology and its clinical relevance," *Br J Anaesth* 111(1) (2013): 13-8.

[9] Rebecca Dale, Brett Stacey, "Multimodal Treatment of Chronic Pain," *Med Clin North Am* 100(1) (2016): 55-64.

[10] Jo Nijs, Oliver Mairesse, Daniel Neu, Laurence Leysen, Lieven Danneels, Barbara Cagnie, Mira Meeus, Maarten Moens, Kelly Ickmans, Dorien Goubert, "Sleep Disturbances in Chronic Pain: Neurobiology, Assessment, and Treatment in Physical Therapist Practice," *Phys Ther* 98(5) (2018): 325-335.

Also, a patient's dignity should be respected.[11] Pain is subjective and should be recognized as such by patients, physicians and nurses. Improving the relationship, communication and trust between patient and doctor is essential for better treatment.

Pain medicine can be demanding, especially in the context of existing economic, official and public environments.[12]

There are many different financial, regional and cultural differences with regard to treating pain. One of them is the lack of identification of pain as a priority in patient care. Another problem is inadequate information about a satisfactory treatment with analgesics.[13] Additionally, problems in treating chronic pain are as follows: undertreatment of pain, inappropriate and excessive consumption of pharmacologic agents, contradictory information between theory and practice, treatment of chronic pain in vulnerable patients such as the elderly.[14]

Pain can exist as an indication of disease and as a phenomenon *per se*.[15] Refractory and chronic pain have complex mechanisms and require diverse specialists, different diagnostic procedures, recordings and treatments. It is an ethical obligation to treat pain. Unfortunately, numerous investigations confirm that pain often goes untreated or undertreated. The American Pain Society presented one study in which 80% of patients after surgical interventions had postoperative pain, with 75% of the patients having moderate, severe or extreme pain.[16] Pain at

---

[11] Karin Fredrikson, Tracy Fasolino, "Pharmacogenetic Testing: The Ethics of Implementing in Clinical Practice for Chronic Pain Patients," *J Pain Palliat Care Pharmacother* 34 (2020): 69-76.

[12] Lilla Náfrádi, Zlatina Kostova, Kent Nakamoto, Peter Schulz, "The doctor-patient relationship and patient resilience in chronic pain: A qualitative approach to patients' perspectives," *Chronic Illn* 14(4) (2018): 256-270.

[13] Ru-Rong Ji, Andrea Nackley,Yul Huh, Niccolo Terrando, William Maixner, "Neuroinflammation and Central Sensitization in Chronic and Widespread Pain," *Anesthesiology* 129(2) (2018): 343-366.

[14] Ana Sofia Carvalho, Sandra Martins Pereira, Antonio Jácomo, Susana Magalhães, Joana Araújo, Pablo Hernández-Marrero, Carlos Costa Gomes, Michael E. Schatman, "Ethical decision making in pain management: a conceptual framework," *J Pain Res* 11 (2018): 967-76.

[15] James Giordano, Michael E Schatman, "An ethical analysis of crisis in chronic pain care. Part 1. Facts, issues, and problems in pain medicine," *Pain Physician* 11(4) (2008): 483–490.

[16] American Pain Society, "Guidelines on the management of postoperative pain," *J Pain* 17(2) (2016): 131–157.

the end of life in patients with different diagnoses and diseases usually remained untreated. In another study, the authors concluded that almost one in two patients with cancer pain was unrelieved and undertreated.[17] The extent to which the problem of chronic pain is expressed in daily work of healthcare professionals is illustrated by the activities of emergency physicians. In total, 4% of all interventions in an emergency unit consisted of calls from cancer patients. In 2009, in Serbia, in one emergency unit 255 interventions per month and 3060 visits per annum were registered because of pain. Most of these interventions were related to the management of exacerbations of oncological disease, where pain was a predominant symptom.[18] Pharmacotherapy in over 75% of cases ensures the elimination of cancer pain when properly administered. [19]

Guidelines for palliative care suggest that adequate pharmacological treatment can eliminate every type of pain in most patients (70-90%). However, everyday practice and research by the World Health Organization show that such treatments, especially in patients suffering from malignant pain, is successful in about 50% of cases. A particular ethical problem is euthanasia in terminal patients.

The treatment of chronic pain is multimodal and requires a multidisciplinary approach that includes different therapeutic modalities. There are guidelines for the treatment and diagnostic techniques for chronic pain. Sound knowledge of pain mechanisms is the basis for an efficient and satisfactory treatment. Physicians with insufficient training cannot help the healthcare system. An interdisciplinary pain management program (PMP) should focus on evidence-based outcomes.[20] Pain clinics offer a biopsychosocial approach to treatment with multidisciplinary PMPs. The aim of these programs is to encourage patients to solve their problems with pain and to move on to a gratifying life in spite of the pain.[21]

---

[17] Robert Swarm, A.M. Abernethy, Judith A. Paice, Doralina L. Anghelescu, et al., "National Comprehensive Cancer Network. Adult Cancer Pain," *J Natl Compr Canc Netw* 11 (8) (2013): 992-1022.

[18] M. Mitrović, B. Jedoksić, S. Petrović, "Opioidi u terapiji hroničnog kancerskog bola," *NČ urgent medic* 16(2) (2010): 78-86.

[19] P. Glare, "Choice of opioids and the WHO Ladder," *J Pediatr Hematol Oncol* 33 (2011): 6-11.

[20] Robert J. Gatchel, Donald D. McGeary, Cindy McGeary, Ben Lippe, "Interdisciplinary chronic pain management: past, present, and future," *Am Psychol* 69(2) (2014): 119-30.

[21] Nicholas Hylands-White, Rui V. Duarte, Jon H. Raphael, "An overview of treatment approaches for chronic pain management," *Rheumatol In* 37 (2017): 29-42.

Analgesics such as non-steroidal anti-inflammatory drugs usually only have a modest effect and significant side effects. Opioids are drugs for the treatment of severe pain conditions. Generally, opioids are safe and effective drugs when used properly. The use of opioid analgesics is a particular ethical problem. Worldwide, the phenomenon of opiophobia is recognized as an irrational fear of using and prescribing opioid analgesics opioids by doctors, nurses, patients and relatives.[22] Opiophobia is an irrational fear of using and prescribing opioid analgesics.

The physicians' fear of using opioids to treat pain is irrational. Therefore, doctors do not prescribe opioids when indicated by the severity of pain, prescribing them at lower doses than necessary or discontinuing them as soon as the pain is resolved, regardless of the cause of the pain. In this way, patients are denied opioids as the most effective and safe medication for the long-term treatment of some types of chronic pain.[23] Insufficient knowledge of pharmacology and pharmacotherapeutic options of opioid analgesics, aided by opiophobia, often leads to inadequate implementation of pain therapy and the use of insufficiently effective drugs.[24] This creates preconditions for poor treatment of patients and inadequate use of all available therapeutic options.

There is also concern about the side effects of opioids. Surveys conducted among physicians show that there is a fear of developing addiction, tolerance and side effects of opioids. Patients often avoid the use of opioids because of fear of tolerance, physical dependence and

---

[22] Ramazan Baldemir, Erkan Yavuz Akçaboy, Özgen Noyan, Zeynep N. Akçaboy, Mustafa Baydar, Şeref Çelik, "An assessment of physicians attitudes toward opioid usage and opiophobia: Results of a survey from a training and research hospital," *Agri* 31 (2019): 23-31.

[23] Andrew Rosenblum, Lisa A. Marsch, Herman Joseph, Russel K. Portenoy, "Opioids and the treatment of chronic pain: controversies, current status, and future directions," *Exp Clin Psychopharmacol* 16 (2008): 405-16. Snežana M. Bosnjak, Martha A. Maurer, Karen M. Ryan, Ivana Popovic, S. Asra Husain, James F. Cleary, Willem Scholten, "A Multifaceted Approach to Improve the Availability and Accessibility of Opioids for the Treatment of Cancer Pain in Serbia: Results From the International Pain Policy Fellowship (2006-2012) and Recommendations for Action," *J Pain Symptom Manage* 52 (2016): 272-83.

[24] Snežana Bošnjak, Dušan B. Beleslin, Ljiljana Vučković-Dekić, "Farmakologija kancerskog bola," *Monografije naučnih skupova AMN SLD* 1 (2007): 55-59.

addiction, respiratory depression, etc. In clinical practice, this means that prescribed opioids are frequently not accepted by patients.[25]

There is also a fear of restrictive regulations that control the prescribing of opioids as narcotics. In practice, an additional problem in the use of these drugs are the legislative and regulative measures for prescribing opioid drugs in different countries.[26]

Opioids have unofficially allowed a place in the treatment of pain. The stigmas and biases associated with opioids have significantly accentuated the barriers of admission, superiority and teaching with regard to appropriate pain management and community debate of potential strategies to promote needed change. Measures that can improve admission, the quality of pain care and resolve public policy debates about the use of opioids for chronic pain are the first steps to ensure an ethically reasonable approach to chronic pain management in the 21st century.

A big problem of multimodal analgesia for chronic pain patients are interactions between drugs. Usually, chronic pain patients have other comorbidities which need appropriate treatment and the use of different drugs, which increases the risk of possible interactions.[27] Interaction can be serious for the patients' health. More importantly, to treat the pain adequately or to treat a serious interaction, represents an ethical issue.

Chronic pain is a significant and costly problem in the United States as well as the rest of the developed world. In the USA, the cost of the treatment of pain is about $65 billion a year. Pharmacotherapy is very important, but a benefit-cost-effectiveness risk should be included in the decision. Planned interdisciplinary programs for the treatment of chronic pain are good and helpful but expensive.

A special ethical problem is treating pain in the elderly.[28] Chronic pain in older adults is connected with special differences, which are age-

---

[25] Michal Graczyk, Alina Borkowska, Molgorzata Krajnik, "Why patients are afraid of opioid analgesics: a study on opioid perception in patients with chronic pain," *Pol Arch Intern Med* 128 (2018): 89-97.

[26] Guy Schofield, Idris Baker, Rachel Bullock, Hannah Clare, Paul Clark, Derek Willis, Craig Gannon, Rob George, "Palliative opioid use, palliative sedation and euthanasia: reaffirming the distinction," *J Med Ethics* 46 (2020): 48-50.

[27] Emmanuel H. During, John W. Winkelman, "Drug Treatment of Restless Legs Syndrome in Older Adults," *Drugs Aging* 39 (2019): 939-946.

[28] Hans Georg Kress, Karsten Ahlbeck, Dominic Aldington, et al, "Managing chronic pain in elderly patients requires a CHANGE of approach," *Curr Med Res Opin* 30 (2014): 1153–64.

related. Pain evaluation is more difficult in old persons. Sometimes, pain can go unreported in the elderly because of dementia and the fear of death.[29] Also, other problems in the elderly are specific changes in pharmacokinetics and pharmacodynamics of certain drugs. The use of opioid drugs in the elderly can be difficult due to interactions with other drugs and comorbidities.[30] There are insufficient clinical studies about pain in the elderly as a vulnerable category.[31] It is a special problem to conduct clinical trials with the older population as vulnerable subjects. Informed consent in clinical trials with the elderly can be problematic due to lack of memory and cognitive impairments such as dementia.[32] This may be one of the reasons why so few randomized, controlled clinical trials on the use of drugs, especially opioids in the elderly, are available. Also, patients that are enrolled in clinical investigations are not representative of most chronic pain patients.[33]

Ethical issues regarding chronic pain are numerous. The barriers to good chronic pain management are complex and multiple: insufficient education, the lack of institutional mechanisms for standardizing the evaluation and treatment of pain, the lack of responsibility for the undertreatment of pain, federal and state permissions for the regulation and adequate use of opioids.[34]

---

[29] B.R. Ferrell, M.T. Eberts, M. McCaffery, M. Grant, "Clinical decision making and pain," *Cancer Nurs* 14 (1991): 289-297.

[30] Milica Prostran, Katarina Savić Vujović, Sonja Vučković, Branislava Medić, Dragana Srebro, Nevena Divac, Radan Stojanović, Aleksandar Vujović, Lepa Jovanović, Ana Jotić, Nataša Cerovac, "Pharmacotherapy of Pain in the Older Population: The Place of Opioids," *Front Aging Neurosci* 8 (2016): 144.

[31] Sebastiano Mercadante, Antonino Giarratano, "Assessing age and gender in studies of breakthrough pain medications," *Curr Med Res Opi*n 30 (2014): 1353–1356.

[32] Milica Prostran, Zoran Todorovic, Radan Stojanovic, Tatjana Potpara, Zorica Nesic, elenaJ Lazic, et al, "Bioethics in clinical trials: vulnerable subjects," in *Bioethics and Pharmacology: Ethics in Preclinical and Clinical Drug Development*, ed. Zoran Todorovic, Milica Prostran, Karel Turza (Trivandrum: Research Network, 2012), 87–100.

[33] William Maixner, Roger B. Fillingim, David A. Williams, Shad B. Smith, Gary D. Slade, "Overlapping Chronic Pain Conditions: Implications for Diagnosis and Classification, " *J Pain* 17 (2016): 93-107.

[34] Volker Tronnier, "Neuromodulation for the Treatment of Chronic Pain - What is Useful? " *Fortschr Neurol Psychiatr* 85 (2017): 223-240.

CONCLUSION

Further clinical studies that examine ethical issues could improve this situation. In an age of rapid medical advances and technological development, it is unacceptable that there is an obstacle to adequate chronic pain medication.

ACKNOWLEDGEMENTS

This work was supported by the Ministry of Education, Science and Technological Development of Serbia (Grant 175023).

REFERENCES

American Pain Society. "Guidelines on the management of postoperative pain." *J Pain* 17(2) (2016): 131–157.

Baldemir, Ramazan, AND Erkan Yavuz Akçaboy, Özgen Noyan, Zeynep N Akçaboy, Mustafa Baydar, Şeref Çelik. "An assessment of physicians attitudes toward opioid usage and opiophobia: Results of a survey from a training and research hospital." *Agri* 31 (2019): 23-31.

Bosnjak, Snežana M., and Martha A. Maurer, Karen M. Ryan, Ivana Popovic, S. Asra Husain, James F. Cleary, Willem Scholten. "A Multifaceted Approach to Improve the Availability and Accessibility of Opioids for the Treatment of Cancer Pain in Serbia: Results from the International Pain Policy Fellowship (2006-2012) and Recommendations for Action." *J Pain Symptom Manage* 52 (2016): 272-283.

Bošnjak, Snežana, and Dušan B. Beleslin, Ljiljana Vučković-Dekić, "Farmakologija kancerskog bola." *Monografije naučnih skupova AMN SLD*, 1 (2007): 55-59.

Buchman, Daniel Z., and Anita Ho, Judy Illes. "You present like a drug addict: patient and clinician perspectives on trust and trustworthiness in chronic pain management." *Pain Med* 17(8) (2016): 1394–1406.

Carvalho, Ana Sofia, and Sandra Martins Pereira, Antonio Jácomo, Susana Magalhães, Joana Araújo, Pablo Hernández-Marrero, Carlos Costa Gomes, Michael E Schatman. "Ethical decision making in pain management: a conceptual framework." *J Pain Res* 11 (2018): 967-976.

Dale, Rebecca, and Brett Stacey. "Multimodal Treatment of Chronic Pain." *Med Clin North Am* 100(1) (2016): 55-64.

During, Emmanuel H., and John W Winkelman. "Drug Treatment of Restless Legs Syndrome in Older Adults." *Drugs Aging* 39 (2019): 939-946.

Ferrell, B.R., and M.T. Eberts, M. McCaffery, M. Grant. "Clinical decision making and pain." *Cancer Nurs* 14 (1991): 289–297.

Fredrikson, Karin, and Tracy Fasolino. "Pharmacogenetic Testing: The Ethics of Implementing in Clinical Practice for Chronic Pain Patients." *J Pain Palliat Care Pharmacother* 34 (2020): 69-76.

Gatchel, Robert J., and Donald D. McGeary, Cindy A. McGeary, Ben Lippe. "Interdisciplinary chronic pain management: past, present, and future." *Am Psychol* 69(2) (2014): 119-30.

Giordano, James, and Michael E. Schatman. "An ethical analysis of crisis in chronic pain care. Part 1. Facts, issues, and problems in pain medicine." *Pain Physician* 11(4) (2008): 483–490.

Glare, Paul. "Choice of opioids and the WHO Ladder." *J Pediatr Hematol Oncol* 33 (2011): 6-11.

Graczyk, Michal, and Alina Borkowska, Molgorzata Krajnik. "Why patients are afraid of opioid analgesics: a study on opioid perception in patients with chronic pain." *Pol Arch Intern Med* 128 (2018): 89-97.

Grichnik, K.P., and F.M. Ferrante. "The difference between acute and chronic pain." *Mt Sinai J Med* 58(3) (1991): 217-20.

Hecke, O. van, and N. Torrance, B.H. Smith. "Chronic pain epidemiology and its clinical relevance." *Br J Anaesth* 111(1) (2013): 13-18.

Hylands-White, Nicholas, and Rui V. Duarte, Jon H. Raphael. "An overview of treatment approaches for chronic pain management." *Rheumatol In* 37 (2017): 29-42.

Ji, Ru-Rong, and Andrea Nackley, Yul Huh, Niccolo Terrando, William Maixner. "Neuroinflammation and Central Sensitization in Chronic and Widespread Pain." *Anesthesiology* 129(2) (2018): 343-366.

Kirsh, K., and A.K. Vice, S. Passik. "History of Opioids and Opiophobia." In *Pain and chemical dependency*, ed. H. Smith, S. Passik, 3-8. New York: Oxford University Press, 2008.

Kress, Hans Georg, and Karsten Ahlbeck, Dominic Aldington, et al. "Managing chronic pain in elderly patients requires a CHANGE of approach." *Curr Med Res Opin* 30 (2014): 1153–1164.

Maixner, William, and Roger B. Fillingim, David A. Williams, Shad B. Smith, Gary D. Slade. "Overlapping Chronic Pain Conditions:

Implications for Diagnosis and Classification." *J Pain* 17 (2016): 93-107.

Mercadante, Sebastiano, and Antonino Giarratano. "Assessing age and gender in studies of breakthrough pain medications." *Curr Med Res Opin* 30 (2014): 1353–1356.

Mitrović, M., and B. Jedoksić, S. Petrović. "Opioidi u terapiji hroničnog kancerskog bola." *NČ urgent medic* 16(2) (2010): 78-86.

Náfrádi, Lilla, and Zlatina Kostova, Kent Nakamoto, Peter Schulz. "The doctor-patient relationship and patient resilience in chronic pain: A qualitative approach to patients' perspectives." *Chronic Illn* 14(4) (2018): 256-270.

Nijs, Jo, and Oliver Mairesse, Daniel Neu, Laurence Leysen, Lieven Danneels, Barbara Cagnie, Mira Meeus, Maarten Moens, Kelly Ickmans, Dorien Goubert. "Sleep Disturbances in Chronic Pain: Neurobiology, Assessment, and Treatment in Physical Therapist Practice." *Phys Ther* 98(5) (2018): 325-335.

Prostran, Milica, and Katarina Savić Vujović, Sonja Vučković, Branislava Medić, Dragana Srebro, Nevena Divac, Radan Stojanović, Aleksandar Vujović, Lepa Jovanović, Ana Jotić, Nataša Cerovac. "Pharmacotherapy of Pain in the Older Population: The Place of Opioids." *Front Aging Neurosci* 8 (2016): 144.

Prostran, Milica, and Zoran Todorovic, Radan Stojanovic, Tatjana Potpara, Zorica Nesic, Jelena Lazic. "Bioethics in clinical trials: vulnerable subjects." In *Bioethics and Pharmacology: Ethics in Preclinical and Clinical Drug Development*, ed. Zoran Todorovic, Milica Prostran, Karel Turza, 87–100. Trivandrum: Research Network, 2012.

Rosenblum, Andrew, and Lisa A. Marsch, Herman Joseph, Russel K. Portenoy. "Opioids and the treatment of chronic pain: controversies, current status, and future directions." *Exp Clin Psychopharmacol* 16 (2008): 405-16.

Schofield, Guy, and Idris Baker, Rachel Bullock, Hannah Clare, Paul Clark, Derek Willis, Craig Gannon, Rob George. "Palliative opioid use, palliative sedation and euthanasia: reaffirming the distinction." *J Med Ethics* 46 (2020): 48-50.

Swarm, Robert, and A.M. Abernethy, Judith A. Paice, Doralina L. Anghelescu, et al., "National Comprehensive Cancer Network. Adult Cancer Pain." *J Natl Compr Canc Netw* 11 (8) (2013): 992-1022.

Treede, Rolf-Detlef. "The International Association for the Study of Pain definition of pain: as valid in 2018 as in 1979, but in need of regularly updated footnotes." *Pain Rep* 3(2) (2018): e643.

Tronnier, Volker. "Neuromodulation for the Treatment of Chronic Pain - What is Useful?" *Fortschr Neurol Psychiatr* 85 (2017): 223-240.

Vucković, Sonja, and Milica Prostran, Milovan Ivanović, Zoran Todorović, Radan Stojanović, Zorica Nesić, Ivana Matić, Slobodan Milovanović. "Opioid analgesics." *Vojnosanit Pregled* 61(4) (2004): 13-21.

THE MEDICALIZATION OF CANNABIS AS AN ETHICAL CHALLENGE IN THE 21st CENTURY

Danijela Vučević,[1] Janko Samardžić,[2] Bojan Jorgačević,[1] Igor Pantić,[3] Jovana Paunović Pantić,[1] Tatjana Radosavljević[1]

INTRODUCTION

> Cannabis is not a medicine. It is poisonous and addictive. This is what we've learned, and we also have our own prejudices, and it's hard, hard to get it out of my head… I prescribe other medications much more freely than I would have prescribed cannabis. That's it. That's a fact. But I am willing to be persuaded that I'm wrong. Especially after I heard Professor Raphi Mechoulam. I heard him several times, two or three times. I was very impressed.
>
> Family physician (physician 19)[4]

Cannabis or hemp plant is an annual flowering plant that grows in tropical and temperate climates, and belongs to the *Cannabis* genus of the

---

[1] Institute of Pathophysiology "Ljubodrag Buba Mihailović," Faculty of Medicine, University of Belgrade, Serbia.

[2] Institute of Pharmacology, Clinical Pharmacology and Toxicology, Faculty of Medicine, University of Belgrade, Serbia.

[3] Laboratory for Cellular Physiology, Institute of Medical Physiology "Rihard Burian," Faculty of Medicine, University of Belgrade, Serbia. University of Haifa, Israel.

[4] Yuval Zolotov, Simon Vulfsons, Dana Zarhin, and Sharon Sznitman, "Medical cannabis: An oxymoron? Physicians' perceptions of medical cannabis," *International Journal of Drug Policy* 57 (2018): 4-10.

family *Cannabinaceae.*[5] With its history dating twelve thousand years back, cannabis is one of the most widely used plant for different traditional healings.[2] However, perhaps no topic is more controversial than the use of compounds and extracts from cannabis in clinical practice, having in mind significant negative effects of smoking marijuana on physical and mental health, as well as social and occupational functioning.[6] Namely, marijuana has a range of adverse health effects, particularly relating to young people, because of higher risk for psychosis, traffic accidents and cognitive impairment.[7] The Centers for Disease Control and Prevention (CDC) refers to marijuana as an illegal drug, and The Drug Enforcement Agency (DEA) classifies marijuana as Schedule I along with heroin and lysergic acid diethylamide.[8]

The psychoactive- and medicinal uses of *Cannabis sativa* (marijuana) have been well known for millennia.[9] The first documentation of

---

[5] Elizabeth Black, Brian Hocum, and Kevin Black, *Ethics and science, cannabinoids and healthcare* (Chicago: Relias Media, 2018). Ran Abuhasira, Liat Shbiro, and Yuval Landschaft, "Medical use of cannabis and cannabinoids containing products - Regulations in Europe and North America," *European Journal of Internal Medicine* 49 (2018): 2-6. Andrea Mastinu, Marika Premoli, Giulia Ferrari-Toninelli, Simone Tambaro, Giuseppina Maccarinelli, Maurizio Memo, and Sara Anna Bonini, "Cannabinoids in health and disease: pharmacological potential in metabolic syndrome and neuroinflammation," *Hormone Molecular Biology and Clinical Investigation* 36 (2018): doi: 10.1515/hmbci-2018-0013. Hemant Goyal, Rubayat Rahman, Abhilash Perisetti, Nihar Shah, and Rajiv Chhabra, "Cannabis in liver disorders: a friend or a foe?," *The European Journal of Gastroenterology & Hepatology* 30 (2018): 1283-1290. Shira Hirsch and Joseph Tam, "*Cannabis*: From a plant that modulates feeding behaviors toward developing selective inhibitors of the peripheral endocannabinoid system for the treatment of obesity and metabolic syndrome," *Toxins* 11 (2019): doi: 10.3390/toxins 11050275.

[6] Elizabeth Black, et al., *Ethics and science, cannabinoids and healthcare*.

[7] Elizabeth Black, et al., *Ethics and science, cannabinoids and healthcare*. Judith Booth and Jörg Bohlmann, "Terpenes in *Cannabis sativa* – from plant genome to humans," *Plant Science* 284 (2019): 67-72. Sabrina Jarvis, Sean Rassmussen, and Blaine Winters, "Role of the endocannabinoid system and medical cannabis," *The Journal for Nurse Practitioners* 13 (2017): 525-531. Maria Steenkamp, Esther Blessing, Isaac Galatzer-Levy, Laura Hollahan, and William Anderson, "Marijuana and other cannabinoids as a treatment for posttraumatic stress disorder: a literature review," *Journal of Depression and Anxiety* 34 (2017): 207-216. Robert Torrence, Don Rojas, and Lucy Troup, "Residual effects of cannabis use on attentional bias towards fearful faces," *Neuropsychologia* 119 (2018): 482-488.

[8] Elizabeth Black, et al., *Ethics and science, cannabinoids and healthcare*.

[9] Joseph Tam, Liad Hinden, Adi Drori, Shiran Udi, Shahar Azar, and Saja Baraghithy, "The therapeutic potential of targeting the peripheral endocannabinoid/CB1 receptor system," *European Journal of Internal Medicine* 49 (2018): 23-29.

cannabis as a medicine appeared in China 5000 years ago when it was recommended for malaria, constipation, rheumatic pains and, mixed with wine, as a surgical analgesic.[10] In India, even before 1000 years BC, this plant was used for various functions, such as hypnotic medicine and a tranquiliser useful in the treatment of anxiety, mania and hysteria.[11] Also, the Assyrians inhaled cannabis to relieve symptoms of depression.[12] Pedacius Dioscorides, a Greek physician, between 50 and 70 AD classified different plants, including *Cannabis sativa*, and described the benefits derived from its use in "*De Materia Medica.*"[13] Roman historian Pliny the Younger, in his pioneering encyclopedia on the natural world, mentions the use of cannabis, too.[14]

Although medicinal preparations containing cannabis were widely used in many societies for centuries, Dr. William O'Shaughnessy introduced it as a modern medicine in Europe in 1839 after the publication of his book "*On the preparations of the Indian hemp, or gunjah.*" [15] Thus, in 1842, he emphasized the efficacy of "Indian hemp," or

[10] Andrea Mastinu et al., "Cannabinoids in health and disease: pharmacological potential in metabolic syndrome and neuroinflammation.'

[11] Raphael Mechoulam, Shimon Ben-Shabat, Lumir Hanus, Moshel Ligumsky, Norbert Kaminski, Antony Schatz, Asher Gopher, Shlomo Almog, Billy Martin, David Compton, Roger Pertwee, Graeme Griffin, Michael Bayewitch, Jacob Barg, and Zvi Vogel, "Identification of an endogenous 2-monoglyceride, present in canine gut, that binds to cannabinoid receptors," *Biochemical Pharmacology* 50 (1995): 83-90.

[12] Ethan Russo and Geoffrey Guy, "A tale of two cannabinoids: the therapeutic rational for combining tetrahydrocannabinol and cannabidiol," *Medical Hypotheses* 66 (2006): 234-246.

[13] Andrea Mastinu et al., "Cannabinoids in health and disease: pharmacological potential in metabolic syndrome and neuroinflammation."

[14] Elizabeth Black et al., *Ethics and science, cannabinoids and healthcare.*

[15] Elizabeth Black et al., *Ethics and science, cannabinoids and healthcare.* Andrea Mastinu et al. "Cannabinoids in health and disease: pharmacological potential in metabolic syndrome and neuroinflammation.' 'Subodh Kumar, Phulen Sarma, Harish Kumar, Ajay Prakash, and Bikash Medhi, "Modulation of endocannabinoid system: Success lies in the failures," *Indian Journal of Pharmacology* 50 (2018): 155-158. William Brooke O'Shaughnessy, "On the preparations of the Indian hemp, or gunjah (*Cannabis indica*): Their effects on the animal system in health, and their utility in the treatment of tetanus and other convulsive diseases," *Transactions of the Medical and Physical Society of Bengal* (1838-1840): 421-461. William Brooke O'Shaughnessy, "On the preparations of the Indian hemp, or gunjah (*Cannabis indica*)," *Provincial Medical Journal and Retrospect of the Medical Sciences* 123 (1843): 363-369. Craig Reinarman, Helen Nunberg, Fran Lanthier, and Tom Heddleston, "Who are medical marijuana patients? Population characteristics from nine California assessment clinics," *Journal of Psychoactive Drugs* 43 (2011): 128-135.

psychoactive cannabis, as a painkiller, a muscle relaxant, and "an anticonvulsive remedy of the greatest value."[16] In Paris, Moreau used cannabis for psychiatric patients and summarized in his published book "*Du Hachisch et de l'Alienation Mentale: Etudes Psychologiques*" describing acute effects of cannabis.[17] Medical interest in cannabis remained in the limelight worldwide for several years due to its psychoactive effects, as well as possible therapeutic use in other domains, such as cholera, rabies and tetanus.[18]

At the beginning of the 20th century, cannabis extracts were used for the treatment of mental disorders, especially as sedatives and hypnotics.[19] Cannabis-based therapeutic agents were prescribed in medical practice in the United States of America (USA) for more than 90 years.[20] Namely, cannabis was listed in the USA pharmacopeia from 1851 to 1944, when it was removed due to fears related to its psychoactive effects.[21] In addition, by the 1930s, cannabis had become subject to federal regulations in the USA, which eventually prohibited its use by the Marijuana Tax Act of 1937.[22] The prohibition was accompanied by moral demonization of cannabis, as it was presented to the public as a harmful drug associated with crime and insanity, so its use decreased rapidly.[23] However, after the identification of the main components of cannabis and the discovery of the endocannabinoid (EC) system as a key part of human physiology, in recent decades cannabis is subject of global

[16] Elizabeth Black et al., *Ethics and science, cannabinoids and healthcare.*

[17] Jacques Joseph Moreau, *Du Hachisch et de l'Alienation Mentale: Etudes Psychologiques, Paris, Librarie de Fortin Mason, 1845* (*English edition*, New York: Raven Press, 1972).

[18] Ibidem.

[19] Ethan Russo and Geoffrey Guy, "A tale of two cannabinoids: the therapeutic rational for combining tetrahydrocannabinol and cannabidiol," 234-246.

[20] Hemant Goyal et al., "Cannabis in liver disorders: a friend or a foe?," 1283-1290.

[21] Elizabeth Black et al, *Ethics and science, cannabinoids and healthcare*. Ethan Russo and Geoffrey Guy, "A tale of two cannabinoids: the therapeutic rational for combining tetrahydrocannabinol and cannabidiol," 234-246. Hemant Goyal et al., "Cannabis in liver disorders: a friend or a foe?," 1283-1290.

[22] Yuval Zolotov et al., "Medical cannabis: An oxymoron? Physicians' perceptions of medical cannabis," 4-10. Daniel Kruger and Jessica Kruger, "Medical cannabis users' comparisons between medical cannabis and mainstream medicine, "*Journal of Psychoactive Drugs* 51 (2019): 31-36.

[23] Yuval Zolotov et al., "Medical cannabis: An oxymoron? Physicians' perceptions of medical cannabis," 4-10. Ran Abuhasira et al., "Medical use of cannabis and cannabinoids containing products," 2-6.

reaffirmation, supported with large number of clinical studies for its medical potential.[24]

Modern era for cannabis started during the 1960s by Mechoulam's group, who isolated and reported the correct structure and stereochemistry of cannabidiol (CBD), Δ^9-tetrahydrocannabinol (Δ^9-THC, the main psychoactive component of marijuana),[25] Δ^8-tetrahydrocannabinol (Δ^8-THC),[26] cannabigerol,[27] cannabichromene (CBC),[28] and cannabicyclol (CBL).[29] Currently, 545 natural compounds have been identified from cannabis.[30] On these, 144 have been isolated and identified as cannabinoids (phytocannabinoids).[31] The first attempt to successfully identify a cannabinoid was made in 1899 by Wood and his colleagues,[32] who isolated cannabinol (CBN).[33]

In the early 1990s, with the isolation of endogenous cannabinoids (endocannabinoids) termed as anandamide (arachidonoylethanolamide/

---

[24] Pál Pacher, Sándor Batkai, and George Kunos, "The endocannabinoid system as an emerging target of pharmacotherapy," *Pharmacological Reviews* 58 (2006): 389-462. Ran Abuhasira et al., "Medical use of cannabis and cannabinoids containing products - Regulations in Europe and North America," 2-6. Andrea Mastinu et al., "Cannabinoids in health and disease." Shira Hirsch and Joseph Tam, "*Cannabis.*" Subodh Kumar et al., "Modulation of endocannabinoid system," 155-158.

[25] Raphael Mechoulam and Youval Shvo, "Hashish. I. The structure of cannabidiol," *Tetrahedron* 19 (1963): 2073-2078. Yehiel Gaoni and Raphael Mechoulam, "Isolation, structure, partial synthesis of an active constituent of hashish," *Journal of American Chemical Society* 86 (1964): 1646-1647. Raphael Mechoulam and Yehiel Gaoni, "The absolute configuration of delta-1-tetrahydrocannabinol, the major active constituent of hashish," *Tetrahedron Letters* 12 (1967): 1109-1111.

[26] Richard Hively, William Mosher, and Fukun Hoffmann, "Isolation of trans-delta-tetrahydrocannabinol from marijuana," *Journal of the American Chemical Society* 88 (1966): 1832-1833.

[27] Yehiel Gaoni and Raphael Mechoulam, "The structure and synthesis of cannabigerol, a new hashish constituent," *Proceedings of the Chemical Society* 86 (1964): 1646-1647.

[28] Yehiel Gaoni and Raphael Mechoulam, "Cannabichromene, a new active principle in hashish," *Chemical Communications* 1 (1966): 20-21.

[29] Leslie Crombie, Roger Ponsford, Arnon Shani, Boris Yagnitinsky, and Raphael Mechoulam, "Hashish components. Photochemical production of cannabicyclol from cannabichromene," *Tetrahedron Letters* 55 (1968): 5771-5772.

[30] Shira Hirsch and Joseph Tam, "*Cannabis.*"

[31] Shira Hirsch and Joseph Tam, "*Cannabis.*" Subodh Kumar et al., "Modulation of endocannabinoid system: Success lies in the failures," 155-158.

[32] Thomas Barlow Wood, Newton Spivey, and Thomas Hill Easterfield, "Cannabinol. Part I," *Journal of the Chemical Society* 75 (1899): 20-36.

[33] Shira Hirsch and Joseph Tam, "*Cannabis.*"

AEA)[34] and 2-arachidonoyl glycerol (2-AG),[35] as well as successful cloning and characterization of cannabinoid-specific receptors (CB1 and CB2) in the brain and peripheral organs,[36] the interest of the scientist community again shifted toward marijuana.[37] The EC system, acting both centrally and peripherally, is ubiquitously present in humans and animals.[38] It has been suggested that its homeostatic roles are "relax, eat, drink, rest, sleep, save, store, forget and protect."[39] However, its role in physiological and pathological conditions are still incompletely known.

---

34 William Devane, Lumir Hanus, Aviva Breuer, Roger Pertwee, Lesley Stevenson, Graeme Griffin, Dan Gibson, Asher Mandelbaum, Alexander Etinger, and Raphael Mechoulam, "Isolation and structure of a brain constituent that binds to the cannabinoid receptor," *Science* 258 (1992): 1946-1949.

35 Takayuki Sugiura et al, "2-arachidonoyl glycerol: A possible endogenous cannabinoid receptor ligand in brain," *Biochemical and Biophysical Research Communications* 215 (1995): 89-97.

36 William Devane, Francis 3rd Dysarz, Ross Johnson, Lawrence Melvin, and Allyn Howlett, "Determination and characterization of a cannabinoid receptor in rat brain," *Molecular Pharmacology* 34 (1988): 605-613. Lisa Matsuda, Stephen Lolait, Michael Brownstein, Alice Young, and Tom Bonner, "Structure of a cannabinoid receptor and functional expression of the cloned cDNA," *Nature* 346 (1990): 561-564. Sean Munro, Kerrie Thomas, and Muna Abu-Shaar, "Molecular characterization of a peripheral receptor for cannabinoids," *Nature* 365 (1993): 61-65. Tian Hua, Kiran Vemuri, Mengchen Pu, Lu Qu, Gye Won Han, Yiran Wu, Suwen Zhao, Wenqing Shui, Shanshan Li, Anisha Korde, Robert Laprairie, Edward Stahl, Jo-Hao Ho, Nikolai Zvonok, Han Zhou, Irina Kufareva, Beili Wu, Qiang Zhao, Michael Hanson, Laura Bohn, Alexandros Makriyannis, Raymond Stevens, and Zhi-Jie Liu, "Crystal structure of the human cannabinoid receptor CB1," *Cell* 167 (2016): 750-762. Zhenhua Shao, Jie Yin, Karen Chapman, Magdalena Grzemska, Lindsay Clark, Junmei Wang, and Daniel Rosenbaum, "High-resolution crystal structure of the human CB1 cannabinoid receptor," *Nature* 540 (2016): 602-606. Tian Hua, Kiran Vemuri, Spyros Nikas, Robert Laprairie, Yiran Wu, Lu Qu, Mengchen Pu, Anisha Korde, Shan Jiang, Jo-Hao Ho, Gye Won Han, Kang Ding, Xuanxuan Li, Haiguang Liu, Michael Hanson, Suwen Zhao, Laura Bohn, Alexandros Makriyannis, Raymond Stevens, and Zhi-Jie Liu, "Crystal structures of agonist-bound human cannabinoid receptor CB1," *Nature* 547 (2017): 468-471. Xiaoting Li, Tian Hua, Kiran Vemuri, Jo-Hao Ho, Yiran Wu, Lijie Wu, Petr Popov, Othman Benchama, Nikolai Zvonok, Kara Locke, Lu Qu, Gye Won Han, Malliga Iyer, Resat Cinar, Nathan Coffey, Jingjing Wang, Meng Wu, Vsevolod Katritch, Suwen Zhao, George Kunos, Laura Bohn, Alexandros Makriyannis, Raymond Stevens, and Zhi-Jie Liu, "Crystal structure of the human cannabinoid receptor CB2," *Cell* 176 (2019): 459-467.

37 Shira Hirsch and Joseph Tam, "*Cannabis*." Subodh Kumar et al., "Modulation of endocannabinoid system," 155-158.

38 Subodh Kumar et al., "Modulation of endocannabinoid system," 155-158.

39 Vincenzo Di Marzo, "Endocannabinoids and other fatty acid derivatives with cannabimimetic properties: Biochemistry and possible physiopathological relevance," *Biochimica et Biophysica Acta* 1392 (1998): 153-175.

In 1999, Health Canada and Canadian Institutes of Health Research jointly started a dedicated Cannabis research initiative named as "Medical Marijuana Research Program."[40] To date, numerous clinical trials are going on evaluating the therapeutic applications of cannabis and its derivatives.[41]

CANNABIS IS A MEDICINE, OR NOT. IS THAT A QUESTION?

Medical cannabis includes raw herbal (botanical) cannabis, cannabis extract and cannabinoids.[42] The taxonomy of cannabis is debated because some people espouse cannabis as a single species with *Cannabis indica*, *Cannabis sativa* and *Cannabis ruderalis* as three different subspecies.[43] On the other hand, some have recommended three distinct species.[44] The term "herbal cannabis" may refer to anyone of these species.[45] Authorization of raw herbal cannabis is still generally unaccepted.[46] Thus, only four countries (Canada, Germany, Israel and the Netherlands), together with more than 50% of the states in the USA have fully authorized its medical use.[47]

Cannabis extract is any extract, usually by organic solvents to produce oil, which is extracted from the plant, and any preparation consisting

[40] Subodh Kumar et al., "Modulation of endocannabinoid system," 155-158.

[41] Subodh Kumar et al., "Modulation of endocannabinoid system," 155-158. Elizabeth Black, et al., *Ethics and science, cannabinoids and healthcare.* Andrea Mastinu et al., "Cannabinoids in health and disease."

[42] Ran Abuhasira et al., "Medical use of cannabis and cannabinoids containing products," 2-6.

[43] Elizabeth Black, Brian Hocum, and Kevin Black, *Ethics and science, cannabinoids and healthcare.* Ran Abuhasira, Liat Shbiro, and Yuval Landschaft, "Medical use of cannabis and cannabinoids containing products," 2-6. Shira Hirsch and Joseph Tam, "*Cannabis.*" Elizabeth Black, Brian Hocum, and Kevin Black, *Ethics and science, cannabinoids and healthcare.* Ran Abuhasira, Liat Shbiro, and Yuval Landschaft, "Medical use of cannabis and cannabinoids containing products," 2-6.

[44] Ran Abuhasira, Liat Shbiro, and Yuval Landschaft, "Medical use of cannabis and cannabinoids containing products," 2-6. Shira Hirsch and Joseph Tam, "*Cannabis*:" Judith Booth and Jörg Bohlmann, "Terpenes in *Cannabis sativa*," 67-72.

[45] Ran Abuhasira, Liat Shbiro, and Yuval Landschaft, "Medical use of cannabis and cannabinoids containing products," 2-6. Shira Hirsch and Joseph Tam, "*Cannabis*:"

[46] Ibidem.

[47] Ran Abuhasira, Liat Shbiro, and Yuval Landschaft, "Medical use of cannabis and cannabinoids containing products," 2-6.

mainly of it.[48] Currently, there are only three cannabinoid-based medicines, available for marketing in different countries[49] (Table 1). These preparations have a statute of borderline products or medicines depending on the concentration of THC in the final product.[50] The good side of cannabinoids is the ability to combine them as an "add-on" therapy with existing conventional therapy.[51]

Table 1. Cannabinoid-based medicine

Medicine and administration route	Content	The main indications and authorization
Nabiximols (Sativex^R) oromucosal spray	cannabinoids THC and CBD	multiple sclerosis associated spasticity EMA* and FDA# not authorized
Nabilone (Cesamet ^{R}) (Canemes^R) oral capsules	Synthetic canabinoid similar to THC	nausea and vomiting due to chemotherapy treatments FDA approved Cesamet in USA
Dronabinol (Marinol^R) (Syndros^R) oral capsules or an oral solution	synthetic Δ^9-THC	anorexia associated with weight loss in AIDS and nausea and vomiting due to cancer chemotherapy FDA approved Dronabinol in USA

* EMA - the European Medicines Agency
# FDA - the United States Food and Drug Administration

---

48 Ibidem.

49 Kiran Vemuri and Alexandros Makriyannis, "Medicinal chemistry of cannabinoids," *Clinical Pharmacology & Therapeutics* 97 (2015): 553-558. Elizabeth Black, Brian Hocum, and Kevin Black, *Ethics and science, cannabinoids and healthcare.* Ran Abuhasira, Liat Shbiro, and Yuval Landschaft, "Medical use of cannabis and cannabinoids containing products," 2-6.

50 Elizabeth Black, et al., *Ethics and science, cannabinoids and healthcare.* Orrin Devinsky, et al., "Trial of cannabidol for drug-resistant seizures in the Dravet syndrome," 2011-2020.

51 Andrea Mastinu, et al., "Cannabinoids in health and disease: pharmacological potential in metabolic syndrome and neuroinflammation." Daniel Kruger and Jessica Kruger, "Medical cannabis users' comparisons between medical cannabis and mainstream medicine,“ 31-36.

Many European countries authorized use of *Nabiximols* and *Nabilone* by the noncentralized route through the competent national authorities.[52] Canada also approved these cannabinoid-based medicines.[53] However, in February 2012, *Dronabinol* was withdrawn from the Canadian market by the manufacturer (not for safety reasons).[54] Besides, there have been considerable arguments in favor of the existence of beneficial cannabinoids, such as CBD.[55] In relation to, EMA has approved CBD for treatment of Dravet syndrome, perinatal asphyxia, graft-versus-host disease, Lennox-Gastaut syndrome, West syndrome and tuberous sclerosis.[56] Additionally, EMA has approved a combination of THC and CBD from extracts of the cannabis sativa for the treatment of glioma.[57] Recently, FDA approved *Epidiolex* (CBD) for the treatment of Lennox-Gastaut syndrome and Dravet syndrome of children.[58] *Epidiolex* is the first FDA-approved drug that contains a purified drug derived from marijuana.[59]

Cannabis use among individuals older than 12 years old has increased from 6.2% in 2002 to 8.3% in 2015,[60] making it one of the most

---

[52] Ran Abuhasira, et al., "Medical use of cannabis and cannabinoids containing products," 2-6.

[53] Orrin Devinsky, Helen Cross, Linda Laux, Eric Marsh, Iin Miller, Rima Nabbout, Ingrid Scheffer, Elizabeth Thiele, Stephen Wright, and Cannabidiol in Dravet Syndrome Study Group, "Trial of cannabidol for drug-resistant seizures in the Dravet syndrome," *The New England Journal of Medicine* 376 (2017): 2011-2020. István Ujváry and Lumir Hanus, "Human metabolites of cannabidiol: a review on their formation, biological activity and relevance in therapy," *Cannabis and Cannabinoid Research* 1 (2016): 90-101. Ran Abuhasira, et al., "Medical use of cannabis and cannabinoids containing products," 2-6. Elizabeth Black, et al., *Ethics and science, cannabinoids and healthcare.*

[54] Ran Abuhasira, et al., "Medical use of cannabis and cannabinoids containing products," 2-6. United Nations Office on Drugs and Crime (UNODC), *World Drug Report 2013* (New York: UNODC, 2013). James Anthony, et al., "Cannabis epidemiology," 6340-6352.

[55] István Ujváry and Lumir Hanus, "Human metabolites of cannabidiol," 90-101.

[56] István Ujváry and Lumir Hanus, "Human metabolites of cannabidiol," 90-101. Elizabeth Black, et al., *Ethics and science, cannabinoids and healthcare.*

[57] Elizabeth Black, et al., *Ethics and science, cannabinoids and healthcare.* Subodh Kumar, et al., "Modulation of endocannabinoid system: Success lies in the failures," 155-158.

[58] Elizabeth Black, et al., *Ethics and science, cannabinoids and healthcare.*

[59] Elizabeth Black, et al., *Ethics and science, cannabinoids and healthcare.* Subodh Kumar, et al., "Modulation of endocannabinoid system: Success lies in the failures," 155-158.

[60] Robert Torrence, et al., "Residual effects of cannabis use on attentional bias towards fearful faces," 482-488.

commonly abused substances in the world after alcohol and tobacco.[61] Nowadays, despite associated health risks with the use of cannabis, its recreational use is permitted in Uruguay and Canada.[62] Simultaneously, medical cannabis is legal in 34 states of the USA.[63] California was beginning a trend toward relaxing legal prohibitions against cannabis production and use.[64] Over the course of several years, other states have passed legislation allowing marijuana purchase, possession and usage to some degree.[65] Today, medical cannabis is also legal in 31 countries (Argentine, Australia, Canada, Chile, Colombia, Croatia, Cyprus /for cancer treatment only/, Czech Republic, Denmark, Finland, Germany, Great Britain, Greece, Israel, Italy, Jamaica, Luxembourg, Malta, Mexico, Netherlands, North Macedonia, Norway, Peru, Poland, Portugal, Republic of Korea, San Marino, Sri Lanka, Switzerland, Uruguay and Zimbabwe).[66] It should be noted that in many countries, there is a considerable gap between the official authorizations, which may be quite permissive, to the actual access of patients to medical cannabis.[67]

Despite the frequent use of the phrase "medicalization of cannabis," it is not clear what it means to different stakeholders (physicians, nurses, patients, employers, insurance companies, pharmaceutical firms, government and regulating bodies, as well as cannabis cultivators and cannabis medicine dispensaries) involved in medical cannabis policy

---

[61] United Nations Office on Drugs and Crime (UNODC), *World Drug Report 2013* (New York: UNODC, 2013). Hemant Goyal, et al., "Cannabis in liver disorders," 1283-1290.

[62] William Tilburg, et al., "Emerging public health law and policy issues concerning state medical cannabis programs," 108-111.

[63] James Anthony, Catalina Lopez-Quintero, and Omayma Alshaarawy, "Cannabis epidemiology: a selective review," *Current Pharmaceutical Design* 22 (2017): 6340-6352. William Tilburg, James Hodge Jr., and Camille Gourdet, "Emerging public health law and policy issues concerning state medical cannabis programs," *The Journal of Law, Medicine & Ethics* 47S2 (2019): 108-111. United Nations Office on Drugs and Crime (UNODC), *World Drug Report 2013* (New York: UNODC, 2013).

[64] Elizabeth Black, et al., *Ethics and science, cannabinoids and healthcare.* Craig Reinarman, e al., "Who are medical marijuana patients?," 128-135.

[65] United Nations Office on Drugs and Crime (UNODC), *World Drug Report 2013* (New York: UNODC, 2013).

[66] William Tilburg, et al., "Emerging public health law and policy issues concerning state medical cannabis programs," 108-111.

[67] Ran Abuhasira, et al., "Medical use of cannabis and cannabinoids containing products," 2-6.

development.[68] Across all different regulatory systems worldwide, physicians hold a vital role in the integration of medical cannabis into clinical practice. As a sociological and theoretical concept, medicalization „describes a process by which nonmedical problems become defined and treated as medical problems, usually in terms of illnesses or disorders."[69] The medicalization or incorporation of cannabis into medicine is complicated, keeping in mind that the knowledge of its properties and effects is still limited.[70] Additionally, cannabis is a plant rather than a pharmaceutical product, which law in many countries following the UN drug convention still defines as an illicit substance with no medicinal value.[71] As the interest in cannabis medical research accelerates (over 10000 scientific papers have been published and its number is rapidly increasing) and the involvement of pharmaceutical companies in developing cannabis-based treatments grows, it is necessary for physicians to be knowledgeable and capable of informing patients on matters related to medicinal or recreational cannabis use.[72]

Biomedical ethics typically recognizes four fundamental principles that have to be weighed against each other in consideration of medical issues from an ethical standpoint.[73] These principles are *beneficence* (the ethical mandate for physicians to help their patients), *respect for autonomy* (the ethical right of every patient to direct his or her own health choices), *non-maleficence* (the ethical mandate for physicians to avoid harm to their

---

[68] Yuval Zolotov, et al., "Medical cannabis: An oxymoron?," 4-10. Subodh Kumar, et al., "Modulation of endocannabinoid system: Success lies in the failures," 155-158. Daniel Kruger and Jessica Kruger, "Medical cannabis users' comparisons between medical cannabis and mainstream medicine," 31-36. Peter Conrad, "Medicalization and social control," 209-232. UN, *Single convention on narcotic drugs*, www.unodc.org/pdf/convention _1961_en.pdf. (accessed February 13, 2021).

[69] Peter Conrad, "Medicalization and social control," *The Annual Review of Sociology* 18 (1992): 209-232. Craig Reinarman, et al., "Who are medical marijuana patients?, 128-135.

[70] UN, *Single convention on narcotic drugs*, www.unodc.org/pdf/convention _1961_en.pdf. (accessed February 13, 2021).

[71] UN, *Single convention on narcotic drugs*, www.unodc.org/pdf/convention _1961_en.pdf. (accessed February 13, 2021). Peter Conrad, "Medicalization and social control," 209-232.

[72] Elizabeth Black, et al., *Ethics and science, cannabinoids and healthcare.* Daniel Kruger and Jessica Kruger, "Medical cannabis users' comparisons between medical cannabis and mainstream medicine," 31-36.

[73] Sylvie Fainzang, "The other side of medicalization: self-medicalization and self medication," *Culture, Medicine and Psychiatry* 37 (2013): 488-504.

patients, or to maximize benefit relative to harm where some degree of harm is unavoidable), and *justice* (the ethical consideration of distribution of health resources in light of relative scarcity, fairness and equality).[74] Although CBD has a positive profile of efficiency and safety, and gives encouraging results in the treatment of patients with inflammations, diabetes mellitus, cancer, affective disorders, neurodegenerative diseases and epilepsy,[75] it is increasingly unethical to continue supporting the traditionally accepted position that there are no legitimate medical applications for cannabis.[76] Studies are continuing to emerge showing compelling benefits in many common pathological and pathophysiological conditions, as well as various functional disorders and organic diseases.[77] In relation to, medical application of cannabis is considered an effective measure in the treatment of pain,[78] anxiety,[79] seizure disorders[80] and multiple sclerosis spasticity.[81] In addition, recently published data indicated the potential usefulness of CB1 blockade in the treatment of high-fat diet (HFD)-induced nonalcoholic fatty liver disease (NAFLD).[82]

---

[74] Elizabeth Black, et al., *Ethics and science, cannabinoids and healthcare.*

[75] Thomas Strouse, "Cannabinoids in medical practice," *Cannabis and Cannabinoid Research* 1 (2016): 38-43.

[76] Elizabeth Black, et al., *Ethics and science, cannabinoids and healthcare.*

[77] Thomas Strouse, "Cannabinoids in medical practice," 38-43.

[78] Thomas Strouse, "Cannabinoids in medical practice," 38-43. Iftach Sagy, et al., "Ethical issues in medical cannabis use," 20-22.

[79] Maria Steenkamp, et al., "Marijuana and other cannabinoids as a treatment for posttraumatic stress disorder: a literature review," 207-216. Robert Torrence, et al., "Residual effects of cannabis use on attentional bias towards fearful faces," 482-488. Ethan Russo and Geoffrey Guy, "A tale of two cannabinoids," 234-246.

[80] Iftach Sagy, Tal Peleg-Sagy, Leonid Barski, Lior Zeller, and Alan Jotkowitz, "Ethical issues in medical cannabis use," *European Journal of Internal Medicine* 49 (2018): 20-22.

[81] Katarina Vesic, Slavica Djukic-Dejanovic, Milica Borovcanin, Janko Samardzic, and Gordana Toncev, "Cannabis as a possible treatment for spasticity in multiple sclerosis," *Serbian Journal of Experimental and Clinical Research* 17 (2016): 61-66. István Ujváry and Lumir Hanus, "Human metabolites of cannabidiol," 90-101. Thomas Strouse, "Cannabinoids in medical practice," 38-43.

[82] Bojan Jorgačević, Dušan Mladenović, Milica Ninković, Milena Vesković, Vesna Dragutinović, Aleksandar Vatazević, Danijela Vučević, Rada Ješić Vukićević, and Tatjana Radosavljević, "Rimonabant improves oxidative/nitrosative stress in mice with nonalcoholic fatty liver disease,"*Oxidative Medicine and Cellular Longevity* (2015): doi:10.1155/2015/842108. Bojan Jorgačević, Danijela Vučević, Ivana Đuričić, Slađana Šobajić, Dušan Mladenović, Milena Vesković, Rada Ješić Vukićević, and Tatjana Radosavljević, "The effect of cannabinoid receptor 1 blockade on hepatic free fatty acid

Generally speaking, physicians in their practice are expected to prescribe unstandardized agents with no EMA and FDA approval with potentially unexpected effects and adverse reactions.[83] Physicians are also required to discuss with their patients' potential risks and benefits before prescribing any treatment or medication.[84] Since medical cannabis dosing and potency is not regulated, there is an unavoidable knowledge gap.[85] Lack of sufficient knowledge regarding the exact content and purity of medical cannabis derivates, and interaction with other drugs impair the patients' ability to reach a fully informed decision. Besides, there is a lack of data on the long term effects of cannabis use, which impacts physician and patient decision making.[86] Moreover, the relationship of some psychiatric conditions with cannabis utilization, such as anxiety, depression, posttraumatic stress disorder, acute psychosis disorders and schizophrenia among users with preexisting genetic vulnerability remained unclear and can put the physician who prescribed them in a constant conflict.[87]

Cannabis has been proposed as an alternative treatment for chronic pain and may help curb the opioid epidemic by serving as a substitute, or as a dose reducer when used in combination with opioids.[88] Therefore allowing cannabis use as a therapeutic approach earlier in the course of the disease may be a more ethical approach for some patients, especially where the risks might be considered low.[89] However, since there is no

---

profile in mice with nonalcoholic fatty liver disease," *Chemistry and Physics of Lipids* 204 (2017): 85-93. Bojan Jorgačević, Danijela Vučević, Milena Vesković, Dušan Mladenović, Dušan Vukićević, Rada Ješić Vukićević, Vera Todorović, and Tatjana Radosavljević, "The effect of cannabinoid receptor 1 blockade on adipokine and proinflammatory cytokine concentration in adipose and hepatic tissue in mice with nonalcoholic fatty liver disease, "*Canadian Journal of Physiology and Pharmacology* 97 (2019): 120-129. Tatjana Radosavljević, "The role of neuroendocrine-inflammatory axis in pathogenesis of nonalcoholic fatty liver disease (in Serbian)," *Medical Investigations (Medicinska Istraživanja)* 51 (2017): 17-18. Danijela Vučević, "The influence of endocannabinoid system on development of metabolic syndrome (in Serbian)" *Medical Investigations (Medicinska Istraživanja)* 52 (2018): 27-28.

[83] Iftach Sagy, et al., "Ethical issues in medical cannabis use," 20-22.

[84] Ibidem.

[85] Ibidem.

[86] Ibidem.

[87] Ibidem.

[88] Ibidem. Elizabeth Black, et al., *Ethics and science, cannabinoids and healthcare*. Ran Abuhasira, et al., "Medical use of cannabis and cannabinoids containing products," 2-6.

[89] Elizabeth Black, et al., *Ethics and science, cannabinoids and healthcare*.

clear guidelines of when to prescribe medical cannabis, the vague indications and relatively high availability of medical cannabis may lead to over-use, misuse, and eventually to illegal trading with third party similar to the broad use of opioids.[90]

Smoking medical cannabis is the most popular administrating method, which may be associated not only with cough, phlegm, bronchitis and other adverse effects on the respiratory tracts of the users, but also with second hand smoke effects on the environment, such as minor impairment on psychomotor abilities and working memory, as well as increase emergency visits and lower cognitive functioning among exposed children.[91] These findings imply that smoked medical cannabis contradicts the harm principle, in which an individual is free to abuse illicit agent unless it does not harm others.[92] On the other hand, according to the principle of respect of autonomy, many patients prefer administrating medical cannabis via smoking methods as the best route of its administrating (shorter half-life, but higher bioviability compared to oral medical cannabis regimens). Thus, these two contradicting ethical considerations need to be balanced.[93]

CONCLUSION

Due to the lack of knowledge on how legalized medical and recreational use of cannabis might affect public health and patients medical choices, more research is needed on this topic. It is increasingly important to educate all stakeholders about its use and side effects, and investigate its potential to be substituted for prescription drugs. Accordingly, physician focus should be on established safety and efficacy data from credible trials and chemically defined compounds, risk vs. benefit concepts, relevant pharmacologic data, and contemporary use trends. Besides, physicians should elicit cannabis use in standard medical histories and consider the pharmacologic implications of cannabis alongside other medications as part of overall medication management. To accomplish this, the use of cannabis should be addressed in a manner that avoids

[90] Iftach Sagy, et al., "Ethical issues in medical cannabis use," 20-22.

[91] Elizabeth Black, et al., *Ethics and science, cannabinoids and healthcare*. Iftach Sagy, et al., "Ethical issues in medical cannabis use," 20-22.

[92] Iftach Sagy, et al., "Ethical issues in medical cannabis use," 20-22.

[93] Ibidem.

bias, presumption, or implicit judgment. Finally, although cannabis is one of the most explored plants in science history, an ethical approach to its medicalization indicates that "conscience without knowledge is not useful, and knowledge without conscience is immorality and danger."

ACKNOWLEDGEMENTS

The Ministry of Education and Science of the Republic of Serbia supported this article through grant N° 175015.

REFERENCES

Abuhasira, R., and Shbiro L., Landschaft Y. "Medical use of cannabis and cannabinoids containing products - Regulations in Europe and North America." *European Journal of Internal Medicine* 49 (2018): 2-6.

Anthony, J., and Lopez-Quintero C., Alshaarawy O. "Cannabis epidemiology: a selective review." *Current Pharmaceutical Design* 22 (2017): 6340-6352.

Black, Elizabeth, and Hocum Brian, Black Kevin. *Ethics and science, cannabinoids and healthcare*. Chicago: Relias Media, 2018.

Booth, J., and Bohlmann J. "Terpenes in *Cannabis sativa* – from plant genome to humans." *Plant Science* 284 (2019): 67-72.

Conrad, P. "Medicalization and social control." *The Annual Review of Sociology* 18 (1992): 209-232.

Crombie, L., and Ponsford R., Shani A., Yagnitinsky B., Mechoulam R. "Hashish components. Photochemical production of cannabicyclol from cannabichromene." *Tetrahedron Letters* 55 (1968): 5771-5772.

Devane, W., and Dysarz F. 3rd, Johnson R., Melvin L., Howlett A. "Determination and characterization of a cannabinoid receptor in rat brain." *Molecular Pharmacology* 34 (1988): 605-613.

Devane, W., and Hanus L., Breuer A., Pertwee R., Stevenson L., Griffin G., Gibson D., Mandelbaum A., Etinger A., Mechoulam R. "Isolation and structure of a brain constituent that binds to the cannabinoid receptor." *Science* 258 (1992): 1946-1949.

Devinsky, O., and Cross H., Laux L., Marsh E., Miller I., Nabbout R., Scheffer I., Thiele E., Wright S., "Cannabidiol in Dravet Syndrome Study Group. Trial of cannabidol for drug-resistant seizures in the Dravet syndrome." *The New England Journal of Medicine* 376 (2017): 2011-2020.

Di Marzo, V. "Endocannabinoids and other fatty acid derivatives with cannabimimetic properties: Biochemistry and possible physio-pathological relevance." *Biochimica et Biophysica Acta* 1392 (1998): 153-175.

Fainzang, S. "The other side of medicalization: self-medicalization and self medication." *Culture, Medicine and Psychiatry* 37 (2013): 488-504.

Gaoni, Y., and Mechoulam R. "Isolation, structure, partial synthesis of an active constituent of hashish." *Journal of the American Chemical Society* 86 (1964): 1646-1647.

——. "The structure and synthesis of cannabigerol, a new hashish constituent." *Proceedings of the Chemical Society* 86 (1964): 1646-1647.

——. "Cannabichromene, a new active principle in hashish." *Chemical Communications* 1 (1966): 20-21.

Goyal, H., and Rahman M., Perisetti A., Shah N., Chhabra R. "Cannabis in liver disorders: a friend or a foe?." *The European Journal of Gastroenterology & Hepatology* 30 (2018): 1283-1290.

Hirsch, S., and Tam J. "*Cannabis*: From a plant that modulates feeding behaviors toward developing selective inhibitors of the peripheral endocannabinoid system for the treatment of obesity and metabolic syndrome." *Toxins* 11 (2019): doi: 10.3390/toxins 11050275.

Hively, R., and Mosher W., Hoffmann F. "Isolation of trans-delta-tetrahydrocannabinol from marijuana." *Journal of the American Chemical Society* 88 (1966): 1832-1833.

Hua, T., and Vemuri K., Pu M., Qu L., Han G. W., Wu Y., Zhao S., Shui W., Li S., Korde A., Laprairie R., Stahl E., Ho J.-H., Zvonok N., Zhou H., Kufareva I., Wu B., Zhao Q., Hanson M., Bohn L., Makriyannis A., Stevens R., Liu Z.-J. "Crystal structure of the human cannabinoid receptor CB1." *Cell* 167 (2016): 750-762.

Hua, T., and Vemuri K., Nikas S., Laprairie R., Wu Y., Qu L., Pu M., Korde A., Jiang S., Ho J.-H., Han G.W., Ding K., Li X., Liu H., Hanson M., Zhao S., Bohn L., Makriyannis A., Stevens R., Liu Z.-J. "Crystal structures of agonist-bound human cannabinoid receptor CB1." *Nature* 547 (2017): 468-471.

Jarvis, S., and Rassmussen S., Winters B. "Role of the endocannabinoid system and medical cannabis." *The Journal for Nurse Practitioners* 13 (2017): 525-531.

Jorgačević, B., and Mladenović D., Ninković M., Vesković M., Dragutinović V., Vatazević A., Vučević D., Ješić Vukićević R.,

Radosavljević T. "Rimonabant improves oxidative/nitrosative stress in mice with nonalcoholic fatty liver disease." *Oxidative Medicine and Cellular Longevity* (2015): doi: 10.1155/2015/842108.

Jorgačević, B., and Vučević D., Đuričić I., Šobajić S., Mladenović D., Vesković M., Ješić Vukićević R., Radosavljević T. "The effect of cannabinoid receptor 1 blockade on hepatic free fatty acid profile in mice with nonalcoholic fatty liver disease." *Chemistry and Physics of Lipids* 204 (2017): 85-93.

Jorgačević, B., and Vučević D., Vesković M., Mladenović D., Vukićević D., Ješić Vukićević R., Todorović V., Radosavljević T. "The effect of cannabinoid receptor 1 blockade on adipokine and proinflammatory cytokine concentration in adipose and hepatic tissue in mice with nonalcoholic fatty liver disease." *Canadian Journal of Physiology and Pharmacology* 97 (2019): 120-129.

Kruger, D., and Kruger J. "Medical cannabis users' comparisons between medical cannabis and mainstream medicine." *Journal of Psychoactive Drugs* 51 (2019): 31-36.

Kumar, S., and Sarma P., Kumar H., Prakash A., Medhi B. "Modulation of endocannabinoid system: Success lies in the failures." *Indian Journal of Pharmacology* 50 (2018): 155-158.

Li, X., and Hua T., Vemuri K., Ho J.-H., Wu Y., Wu L., Popov P., Benchama O., Zvonok N., Locke K., Qu L., Han G.W., Iyer M., Cinar R., Coffey N., Wang J., Wu M., Katritch V., Zhao S., Kunos G., Bohn L., Makriyannis A., Stevens R., Liu Z.-J. "Crystal structure of the human cannabinoid receptor CB2." *Cell* 176 (2019): 459-467.

Mastinu, A., and Premoli M., Ferrari-Toninelli G., Tambaro S., Maccarinelli G., Memo M., Bonini S.A. "Cannabinoids in health and disease: pharmacological potential in metabolic syndrome and neuroinflammation." *Hormone Molecular Biology and Clinical Investigation* 36 (2018): doi: 10.1515/hmbci-2018-0013.

Matsuda, L., and Lolait S., Brownstein M., Young A., Bonner T. "Structure of a cannabinoid receptor and functional expression of the cloned cDNA." *Nature* 346 (1990): 561-564.

Mechoulam, R., and Shvo Y. "Hashish. I. The structure of cannabidiol." *Tetrahedron* 19 (1963): 2073-2078.

Mechoulam, R., and Gaoni Y. "The absolute configuration of delta-1-tetrahydrocannabinol, the major active constituent of hashish." *Tetrahedron Letters* 12 (1967): 1109-1111.

Mechoulam, R., and Ben-Shabat S., Hanus L., Ligumsky M., Kaminski N., Schatz A., Gopher A., Almog S., Martin B., Compton D., Pertwee R., Griffin G., Bayewitch M., Barg J., Vogel Z. "Identification of an endogenous 2-monoglyceride, present in canine gut, that binds to cannabinoid receptors." *Biochemical Pharmacology* 50 (1995): 83-90.

Moreau, Jacques Joseph. *Du Hachisch et de l'Alienation Mentale: Etudes Psychologiques, Paris, Librarie de Fortin Mason 1845*. English edition, New York: Raven Press, 1972.

Munro, S., and Thomas K., Abu-Shaar M. "Molecular characterization of a peripheral receptor for cannabinoids." *Nature* 365 (1993): 61-65.

O'Shaughnessy, W.B. "On the preparations of the Indian hemp, or gunjah (*Cannabis indica*): Their effects on the animal system in health, and their utility in the treatment of tetanus and other convulsive diseases." *Transactions of the Medical and Physical Society of Bengal* (1838-1840): 421-461.

—. "On the preparations of the Indian hemp, or gunjah (*Cannabis indica*)." *Provincial Medical Journal and Retrospect of the Medical Sciences* 123 (1843): 363-369.

Pacher, P., and Batkai S., Kunos G. "The endocannabinoid system as an emerging target of pharmacotherapy." *Pharmacological Reviews* 58 (2006): 389-462.

Radosavljević, T. "The role of neuroendocrine-inflammatory axis in pathogenesis of nonalcoholic fatty liver disease (in Serbian)." *Medical Investigations (Medicinska Istraživanja)* 51 (2017): 17-18.

Reinarman, C., and Nunberg H., Lanthier F., Heddleston T. "Who are medical marijuana patients? Population characteristics from nine California assessment clinics." *Journal of Psychoactive Drugs* 43 (2011): 128-135.

Russo, E., and Guy G. "A tale of two cannabinoids: the therapeutic rational for combining tetrahydrocannabinol and cannabidiol." *Medical Hypotheses* 66 (2006): 234-246.

Sagy, I., and Peleg-Sagy T., Barski L., Zeller L., Jotkowitz A. "Ethical issues in medical cannabis use." *European Journal of Internal Medicine* 49 (2018): 20-22.

Shao, Z., and Yin J., Chapman K., Grzemska M., Clark L., Wang J., Rosenbaum D. "High-resolution crystal structure of the human CB1 cannabinoid receptor." *Nature* 540 (2016): 602-606.

Steenkamp, M., and Blessing E., Galatzer-Levy I., Hollahan L., Anderson W. "Marijuana and other cannabinoids as a treatment for posttraumatic stress disorder: a literature review." *Journal of Depression and Anxiety* 34 (2017): 207-216.

Strouse, T. "Cannabinoids in medical practice." *Cannabis and Cannabinoid Research* 1 (2016): 38-43.

Sugiura, T., and Kondo S., Sukagawa A., Nakane S., Shinoda A., Itoh K., Yamashita A., Waku K. "2-arachidonoyl glycerol: A possible endogenous cannabinoid receptor ligand in brain." *Biochemical and Biophysical Research Communications* 215 (1995): 89-97.

Tam, J., and Hinden L., Drori A., Udi S., Azar S., Baraghithy S. "The therapeutic potential of targeting the peripheral endocannabinoid/ CB1 receptor system." *European Journal of Internal Medicine* 49 (2018): 23-29.

Tilburg, W., and Hodge Jr. J., Camille Gourdet C. "Emerging public health law and policy issues concerning state medical cannabis programs." *The Journal of Law, Medicine & Ethics* 47S2 (2019): 108-111.

Torrence, R., and Rojas D., Troup L. "Residual effects of cannabis use on attentional bias towards fearful faces." *Neuropsychologia* 119 (2018): 482-488.

Ujváry, I., and Hanus L. "Human metabolites of cannabidiol: a review on their formation, biological activity and relevance in therapy." *Cannabis and Cannabinoid Research* 1 (2016): 90-101.

UN. *Single convention on narcotic drugs*. www.unodc.org/pdf/convention _1961_en.pdf. Accessed February 13, 2021.

United Nations Office on Drugs and Crime (UNODC). *World Drug Report 2013*. New York: UNODC, 2013.

Vemuri, K., and Makriyannis A. "Medicinal chemistry of cannabinoids." *Clinical Pharmacology & Therapeutics* 97 (2015): 553-558.

Vesic, K., and Djukic-Dejanovic S., Borovcanin M., Samardzic J., Toncev G. "Cannabis as a possible treatment for spasticity in multiple sclerosis." *Serbian Journal of Experimental and Clinical Research* 17 (2016): 61-66.

Vučević, D. "The influence of endocannabinoid system on development of metabolic syndrome (in Serbian)." *Medical Investigations (Medicinska Istraživanja)* 52 (2018): 27-28.

Wood, T.B., and Spivey N., Easterfield T.H. "Cannabinol. Part I." *Journal of the Chemical Society* 75 (1899): 20-36.

Zolotov, Y., and Vulfsons S., Zarhin D., Sznitman S. "Medical cannabis: An oxymoron? Physicians' perceptions of medical cannabis." *International Journal of Drug Policy* 57 (2018): 4-10.

IS THERE A MORAL REPRODUCTIVE BEHAVIOUR IN THE CONTEXT OF PRENATAL DIAGNOSIS AND CARE?

Luboslava Kostova[1]

INTRODUCTION

Prenatal genetic testing and screening during pregnancy are free and non-mandatory options for women. However, they are almost immediately accepted by every pregnant woman as something good, useful and obligatory in the sense of responsible parenthood and as a source of excellent prenatal care. When the offer of testing is made, the parents` opportunity to reject it is almost inexistent. It is nearly impossible to choose not to choose because this means that you are irresponsible, overly religious or out of date, "crazy" or just a "gypsy" – which is the most common offence in the Bulgarian reproductive discourse. Gravitating towards the concept of moral reproductive behaviour, we gravitate towards the concept of gendered moral responsibility. Mothers are the main actors and their basic role as givers and caretakers is used in constructing this kind of biosocial normality. The choice is theirs, as well as the responsibility and the shame in case of rejecting testing and eventually giving birth to a sick or disabled child. There is a basic consensus that an illness or disability is wrong, something that burdens and not enriches the life of the family. In the context of prenatal testing and screening, the selective abortion of an impaired foetus in case of positive test result and diagnosis is considered not only as the only

[1] Ethical Studies Department, Institute of Philosophy and Sociology – BAS.

option, but also as a responsible and reasonable option, as a right of choice.

Current debates in Poland show this construction very clearly. "The governing right-wing Law and Justice party (PiS) to approve draft legislation that would ban pregnancy terminations in the case of foetuses with congenital birth defects. The worst-case scenario that could come true has come true. It is a devastating sentence that will destroy the lives of many women and families," said Kamila Ferenc, a lawyer who works with an NGO helping women denied abortion.[2] Is life with birth defects always a devastating sentence or is it only our perception, prejudice or our imagination's fail? Who has the right to interpret the reproductive power of a woman's body, the woman herself or the whole society, wombed in her body – state, church, men, and finally her own unborn children. According to Barbara Catz Rothman, what is considered to be a goal in good prenatal care is "not just the making of babies, but the making of mothers that midwives see as the miracle of birth."[3] Therefore, does much knowledge bring great sadness? What can be a good pregnancy outcome and what is the essential basis for wholesomeness or integrity reconsidered today as moral responsibility?

There are at least two ways to understand the midwives' valuing a pregnancy as "good" when the outcome is "bad." The first reflects their view of life. If life is about accomplishing things, then pregnancies resulting in dead babies are pointless. But if life is about living, and we have only limited time to live, then days spent in joyous anticipation are good days, and days spent in grief are bad days. Therefore, the prenatal diagnosis of conditions that cause death simply moves days from the good to the bad side of the ledger for women. A second way of understanding this – closely related to the first – is to postulate that pregnancy itself has a particular meaning and value in a woman's life, and that for women who want to become mothers, a good pregnancy and good birth are good things to have. Consider the following: in our practice, we had a child with a disorder that was not compatible with life.

[2] "Poland rules abortion due to foetal defects unconstitutional," *The Guardian* (2020), www.theguardian.com/world/2020/oct/22/poland-rules-abortion-due-to-foetal-defects-unconstitutional?CMP=share_btn_fb&fbclid=IwAR2d_qvzM_iZUM5CN0j8B4H2xf-I1H9QMeZx9YT709Q9ljQGqCPBXu2VrfU (accessed November 17, 2020).
[3] Barbara Katz Rothman, "Spoiling the pregnancy. Prenatal diagnosis in the Netherlands," in *Birth by design* (London: Routledge, 2001).

It did not have a *middenrif* (diaphragm). The intestines were placed upwards, while the heart was in the wrong place. The mother had a good pregnancy, a difficult delivery, but she looks back on it very positively. The child lived for a couple of hours. Of course, the parents were in grief for the child, but they also had very positive feelings toward the child. I heard several pregnant women talk about it, saying condescendingly, "They could have seen it [the disorder] in an echo" (i.e., an ultrasound), which is true. "They should have done an echo, then they could have known."

This is the general view on the issue: something that is best to be known in advance. Theoretically, however, the women who talked about this mother did not know her. This woman was delighted that she did not know anything in advance because an ultrasound would not have changed it. Yes, she would probably have had a hospital birth and three thousand echoes and pressure. However, she now has the memory of a good pregnancy and a positive experience of her delivery. The outcome would have been the same in both cases."[4]

THE MEANING OF GENDER

To explore the meaning of gender, we have to clarify its existing definitions. According to one of them, gender is a social construct, a social role and social expectation for proper behaviour, rather than a biological destiny. This is a position, which plays a significant role in the deconspiration and demystifying of the structures of power in the social world and also has an obvious weakness, provoked by the concept`s overexposure. The mutual connection between nature and culture is a fact of great significance to the understanding of the human situation: we are neither natural nor cultural beings, but both their synthesis: that allows us to survive.[5]

According to the other theory, sex is a natural, biological construct, something which fits the social order in a restricted and limited way: "It is almost impossible to transform from a man into a woman."[6] The disadvantages of that theory come from two main points: first, it generalizes and reduces the individual's personality only to its biology

---

[4] Ibidem.

[5] T. Eagleton, *The idea of culture* (Hoboken: Wiley, 2000), 131.

[6] Peter Berger, *Invitation to sociology. A Humanistic Perspective* (London: Anchor, 1999), 15.

and second, it attributes social power dimensions and current social inequalities to pure biology (as a destiny). What is typical between these two theories of understanding gender is that they recognize its significance and presence in almost every area of human existence because gender is a structure that influences social and cultural practices.

This article will put the concept of gender in relation to the medicalization of the reproduction process and its direct influence on the prospective parents' decisions during the offer of prenatal diagnosis. The practice of prenatal diagnosis during pregnancy gives parents information about their foetus' health, so in case of detected anomalies such as Down syndrome, they have to decide whether to continue the pregnancy or not. This moment of choice has at least two faces. On the one hand, it is voluntary and free from the interference of physicians and the state. On the other hand, including prenatal genetic testing in national screening programs makes them routine and standard decisions. Converting prenatal diagnosis into a popular mass-decision is legitimated by the ideas of risk and responsibility, health and illness, good and evil, from life worthy and unworthy of living. We become witnesses to tremendous transformations in what we consider to be responsible reproductive behaviour. The scope of testing increases because it is defined as a parents' right, not as an obligation. So the practice of prenatal diagnosis as a part of the genetic testing itself reveals its main characteristics through the control of reproduction.

Reproduction is a substantial and frequently latent part of biopolitical processes. It takes part in developing national policies and also has global dimensions: these aspects of Foucault's works are an object of growing attention in the field of feminist bioethics studies and in disability studies.[7]

Prenatal genetic testing (both invasive and non-invasive procedures) possesses three main biopolitical characteristics, which directly influence the capacity of free and autonomous choice:

- They are free to all parents, mothers to be, who want to use them;
- They treat all pregnancies as risky;
- They assume different discourses in the heart of which are future parents, especially the future mother – a subject of choice and an

[7] Penelope Deutscher, „The Inversion of Exceptionality: Foucault, Agamben, and 'Reproductive Rights'," *South Atlantic Quarterly* 107 (2008): 55-70.

object of governance in the same capacity of choice of taking or not prenatal testing.

THE MEDICALIZATION OF REPRODUCTION

The process of medicalization radically transforms reproduction from actual experience and capacity for autonomy, affecting women's bodies and reproductive functions to means of social control. From an intimate and private sphere, reproduction turns to a delicate and specific interaction between gender, sexuality, politics, economics, biotechnology and ethical dilemmas. The process of medicalization of reproduction reveals global changes, provoked by the status of biomedicine as a legitimate source of truth about the body. In the nineteenth century, building the national state was accompanied by the care of the national population and it gives different roles to men and women and different consequences for each of them.

The new demographic discourse was concerned with both increasing the number of national citizens and improving their overall "quality" in terms of health, productivity, and reproduction. Within this discourse, men's citizenship came to be defined largely in terms of their contributions as disciplined and conscientious soldiers and productive and compliant workers. Women, on the other hand, were assigned the primary responsibility for reproducing and ensuring the health and vigor of these "stalwart sons" of the nation.[8]

Over time, the ideological bond between women and motherhood and between women's citizenship and the quality of their reproductive contribution to the national body became increasingly tight. The new concern with the quantity and quality of national populations was accompanied by the "discovery" of children and their migration from the ideological margins to the center of concern in family life.[9]

By the latter part of the nineteenth century, these newly devised statistical and survey techniques were revealing declining fertility and

[8] Kalpana Ram and Margaret Jolly, *Maternities and Modernities: Colonial and Postcolonial Experiences in Asia and the Pacific* (Cambridge: Cambridge University Press, 1998).
[9] Phillipe Aries, *Centuries of Childhood* (London: Jonathan Cape, 1962); Jacques Donzelot, *The Policing of Families* (New York: Pantheon, 1979).

alarmingly high rates of infant and child mortality in nation after nation across Europe and North America.[10]

More generally, the high rates of infant and child deaths, as well as the declining birth rates, uncovered by the new methods of statistical analysis, were widely read as an abdication on the part of women to do their duty as citizens, either by not reproducing in sufficiently large numbers or by contributing to the premature death of their children through ignorance and the pursuit of irrational childbirth and child-rearing practices. In either case, the stage was set for a series of interventions whose objective was to medicalize motherhood, that is, to replace existing maternal understandings and practices with novel ones based on science and medicine. This was a project in which medicine and its auxiliary professions, as well as education, the law, and later on, the mass media, all played pivotal roles. Yet across the board, it was women who were assigned the core responsibility for learning and following the canons of scientific hygiene and medicine to ensure the successful birth, survival, and health of their children.[11]

"Good mothers" were those who did so, "bad mothers" those who did not. And since all women were expected to reproduce, their contributions to society increasingly came to be gauged by their "successes and failures as mothers."[12]

Medicalized definitions and categories thus had a powerful role to play in reconfiguring gender roles and meanings in the nineteenth and twentieth centuries. Ultimately, however, rather than unifying women in a joint project of scientific reproduction, they offered a new (and constantly mutating) set of tools for dividing, condemning, and regulating them.[13]

Either way, their reactions and responses as users, consumers, and, above all, as subjects of medical discourse and practice are barely audible.

---

[10] Leon Blum, *At the Breast: Ideologies of Breastfeeding and Motherhood in the Contemporary United States* (Boston: Beacon Press, 1999); J. Cole, *The Power of Large Numbers: Population, Politics, and Gender in Nineteenth Century France* (Ithaca and London: Cornell University Press, 2000), 15.

[11] Rima Apple, *Mothers and Medicine: A Social History of Infant Feeding, 1890–1950* (London and Madison: University of Wisconsin Press, 1987).

[12] Joshua Cole, *The Power of Large Numbers*, 15.

[13] Jacqueline Litt, *Medicalized Motherhood: Perspectives from the Lives of African-American and Jewish Women* (New Brunswick, N.J., and London: Rutgers University Press, 2000).

However, a rapidly growing number of feminist-informed ethnographic studies persuasively argue the inadequacy of such approaches and underscores both the scholarly and practical benefits of paying close attention to women's embodied experiences as they attempt to interpret and assign meaning to an ever-expanding array of reproductive technologies and practices.[14]

Current studies suggest that significant cultural differences between the scope and applications of different biomedical techniques and methods in the field of reproduction. "More than any other area of medical practice, the organization and provision of maternity care is a highly charged mix of medical science, cultural ideas and structural forces."[15]

So one of the problems which feminist bioethics deals with is the identification of the woman only as a mother and valuating her duty to care and to serve. The woman is the biological sex, who becomes pregnant, gives birth and raises the children. The new reproductive technologies make it possible to distinguish sexuality from reproduction. The sexual act becomes unnecessary for the human's reproduction. According to feminist argumentation, personal life and experience rather than abstract, moral reasoning is at the core of moral decisions and actions. The area of research of reproductive practices is located between public and private domain where the women's reproductive choices have been made. The concept of individual choice, even made as a choice of concrete medical behaviour and services, is always a part of some greater structural space. In the context of reproduction, these are the discourses of gynaecology, midwifery, demography and the idea of rights.

LIMITS OF CHOICE IN THE DECISION-MAKING PROCESS DURING PRENATAL GENETIC TESTING

Prenatal genetic testing is a part of biotechnological development and its integration into the reproduction medicalization process. In a way, biotechnologies are not only a means of increasing the control of human life but rather a tool to decrease the amount of suffering in it. According

[14] Margaret Lock and Patricia Kaufert, ed., *Pragmatic Women and Body Politics* (Cambridge: Cambridge University Press, 1998).
[15] Raymond De Vries., ed., *Birth by Design: Pregnancy, Maternity Care, and Midwifery in North America and Europe* (New York: Routledge, 2001).

to Freud, suffering has at least three sources: the power of nature, the weakness of the body and the insufficiency of regulations in family, state and society affairs.[16]

These assumptions lead up to the fact that every technology or practice of decreasing the suffering has to be accepted as an appropriate one. The person's choice is to define the limits of such interruption and this is the main ethical question: what will be its scope? The practice of prenatal diagnosis is not only a value-free application of some technology and free exercise of informed and rational choice. Its own existence influences the choice, making it impossible to refuse it. The choice, provoked by the information derived from the prenatal testing, is, nevertheless, action. Alternatives like not choosing, accepting things as they are, refusing such information are defined as a form of irresponsibility. So, on one side, the choice of prenatal testing has medical aspects, and on the other –moral ones, because the decision not to choose has to be justified.[17]

Variety of choices implies more significant responsibilities and technology demands and receives answers through people's choices. Because of the biotechnological implications, our knowledge and imagination related to the hidden foetus increases. The price of this knowledge reformulates the concept of parental care into quality control. In the past, when these tests did not exist, mothers used to interrupt pregnancy because of their situation and life plans and not due to the expected quality of life of the unborn child. When a mother is provided with such biomedical information, this transforms her into a moral pioneer who has to decide whether to allow her child into the world.[18]

THE DECISION FOR PRENATAL GENETIC TESTING

The choice of pregnant women to accept or refuse prenatal testing, to continue or to terminate the wanted pregnancy, emphasizes the meaning of disease and disability to life. To do prenatal testing means to be ready to accept a considerable amount of ethical dilemmas and personal risks

[16] Zigmund Freud, *Civilization and its Discontents* (London: Hogarth, 1975), 57-146.

[17] Mishelle Callon and Vololona Rabeharisoa, "Gino's Lesson on Humanity: Genetics, Mutual Entanglements and the Sociologist's Role," *Economy and Society* 33 (2004): 1-27.

[18] Rayna Rapp, *Testing Women, Testing the Fetus: the Social Impact of Amniocentesis in America* (London: Routledge, 1999), 3.

connected to such a decision. Reasons for accepting prenatal testing could be defined as purely medical, but the realization of such a decision is highly subjective. Research in the field of "risk" has shown significant differences between objective and subjective perception of risk because the latter depends on personality, level of information and scale of values, especially the value of good. As much as the idea of good is of value for a person, they will want to protect it, even from the minor harms and risks. Furthermore, what is important here is that every component of the prenatal testing (invasive and non-invasive) procedures could be a base for a dilemmatic situation.

The purpose of testing is to provide future parents with information about foetal health and detect potential anomalies indirectly so that future parents can be reassured of their child's good health condition. Most medical practices make a synthesis between diagnosis and therapy. A good diagnosis defines successful treatment and vice versa. In the implementation of prenatal testing, this connection is very controversial. They do not function as means of therapy because there is no cure for most of the detected anomalies. In fact, prenatal testing gives information, not therapy. Basically, because of this information, parents become responsible for the life or the death of their child and its eventual suffering if they decide to continue the pregnancy despite the detected disability.

The link between diagnosis, therapy and recovery is broken so the person's mind becomes a source of intense pain, confusion and anxiety. Overcoming this emotional state of shock is a task of great importance and considerable effort. Suffering has to be integrated into person's life and sense of authenticity. Separating the diagnosis from the therapy creates substantial moral conflicts and dilemmas because it happens during pregnancy. Moreover, pregnancy is an intimate experience between the mother and the child (foetus). Through prenatal genetic testing, the right to life of a child (foetus) is opposed to the mother's right of choice.[19]

The parental choice could be separated into two sides: deciding on whether doing prenatal testing at all, and what to do in case of a detected anomaly. Even though most test results are good and indicate the future

[19] Marianne Leuzinger-Bohleber, Eva Engels M., Tsiantis, J., ed., *The Janus Face of Prenatal Diagnostics – a European study bridging ethics, psychoanalysis and medicine* (London: Karnac Books, 2008).

child's health (97% of results), in the cases of a detected anomaly, the situation changes and so does the whole pregnancy. According to data collected by an EDIG study, the main factors for undergoing prenatal testing are maternal age and the results from the triple screen, which is a test for risk indication. The decision for testing is made quickly after a discussion with the physician and the partner with some cross-cultural differences. In Israel, for example, pregnant women make the decision by themselves, while in Greece, the role and authority of the physician is dominating. According to data collected in Germany, only 4% of the population would not undergo prenatal testing and this justifies the practice as a common decision during pregnancy.

Empirical profile of participants in prenatal genetic testing (EDIG, 2008)

Table 1. Reasons for undergoing prenatal testing

Reasons	Germany	Italy	Israel	Greece
Age	61.3 %	68.6 %	48 %	54.6 %
Genetic	8.3 %	3.6 %	15 %	8.8 %
Psychological	7.5 %	2.4 %	0.8 %	2.4 %
Risk assessment test result	13.3 %	12.1 %	23.4 %	17.6 %
Others	9.7 %	12.9 %	11.9 %	16.6 %

Table 2. How has the decision for prenatal testing been made?

Ways	Germany	Israel	Greece	Italy
Discuss only with my physician	18.7 %	14.2 %	28.8 %	16.9 %
Discuss with my partner	68.7 %	62.5 %	60.4 %	71. 2 %
Discuss with relatives	1.5 %	0 %	0.9 %	0.8 %
Discuss with both physician and partner	44 %	35 %	47 %	36 %
Discuss with physician, partner and relatives	12.5 %	10.6 %	7.2 %	5.9 %

Table 3. What kind of choice have you made?

Answers	Germany	Israel	Greece	Italy
Quick	74. 9 %	81.8 %	81.4 %	71.8 %
Complicated	16.5 %	12.9 %	4.7 %	25.4 %

Answers	Germany	Israel	Greece	Italy
Difficult	19.8 %	19. 5 %	18.6 %	13.6 %

Table 4. Is the prenatal testing-decision just yours?

Answers	Germany	Israel	Greece	Italy
It is my decision	16.4 %	27.7 %	11.7 %	9.3 %
It is our decision (me and my partner)	57 %	61.6 %	38.7 %	61.9 %
It is my physician's decision	1.8 %	0.9 %	1.8 %	0.8 %
It is our decision (me and my physician)	11.9 %	10.7 %	25.2 %	5.9 %
It is our decision (me, my partner and my physician)	25.1%	21.4 %	41.4 %	22.9 %

This empirical data illustrates that subtle, but important differences exist between different countries and that the same biomedical practices are accepted elsewhere in different, contextual ways.

THE DECISION AGAINST PRENATAL GENETIC TESTING

The decision against prenatal genetic testing could also be a part of the ethical dilemma. The very existence of the opportunity to test prenatally changes the moral and social implications of disability, even in its later appearance. For example, if prospective parents (namely the future mother) refuse to undergo such testing for Tay-Sacks disease and their baby has that condition, their birth will not be accepted as a tragic state of the art, but rather as a result of conscious and wilful ignorance.

There is a discreet but significant difference between sympathy in the case of a tragic, accidental life event compared to something avoidable. Such testing changes the shape of human solidarity, turning reproduction into an area of individual responsibility. The possibility to explain suffering or accident by chance is very restricted. The concept of individual choice and responsibility in the context of prenatal genetic testing changes not only the total amount of children born with disabilities but also the individual's perceptions of the role of suffering in our lives, transforming it into a rational calculation. The concept of choice is a factor of great importance to the distribution and scope of prenatal testing.[20]

This choice, however, is located into a structurally- defined area of technological development, and it functions by itself rather than as a proof of individual freedom and autonomy. Moreover, this right of choice restricts the right of life, which becomes a subject of selection. In this context, the very structure of non-invasive procedures like the so-called risk-assessment triple functions not only as a way of optimizing prenatal testing, but as a normative artefact. The incorporative normativity influences our space for choice and our actions, when they are based on selective criteria. It embodies the connection between technical availability for anomalies' identification with the normative desire to prevent births of disabled children. This desire is driven by one of the basic principles in medical practice: the principle to avoid harm. For example, the desire to prevent the birth of a child with Down's syndrome is incorporated in the same methods used for its detection. So when the triple test is offered to the pregnant woman in maternity care services, this is a good example of an incorporative normativity, because the offer begins to function by itself, independently of the woman's decision whether to undergo testing or not.[21]

Primarily because of the provided information, the pregnant woman is responsible for making a choice. Even when her decision is not to undergo testing, it has already been an incorporated decision. Thereby, the increase of options leads to a decrease in the capacity to act independently. Every part of the decision-making process is rooted in the concept of benevolence for the child's sake – even the choice of selected abortion in the cases of detected foetal anomaly.

It is worth mentioning that the options for choice express a wrong dichotomy, located in the very heart of the medical model of reproduction: the choice between the desire for a healthy child and an opportunity of the woman to exercise her own autonomy not only to her body but also to her identity as a whole. This dichotomy is a modification

---

[20] Cathrine Mills, "Resisting Biopolitics, Resisting Freedom: Prenatal Testing and Choice," in *Resisting Biopolitics: Philosophical, Political and Performative Strategies*, ed. Wilmer, Stephen & Zukauskaite, Audrone (London: Routledge. 2015), 19.

[21] Marcus Popkema and Hans Harbers, "The Cultural Politics of Prenatal Screening," in *Inside the Politics of Technology: Agency and Normativity in the Co-Production of Technology and Society* (Amsterdam: Amsterdam University Press, 2005), 238.

of the Cartesian dualistic view of body and soul distinction and about the need to rationalize and discipline the latter.

PRENATAL GENETIC TESTING IMPLICATIONS: FOR PREGNANCY, MOTHERHOOD AND PARENTHOOD

In her article "How is technology changing the meaning of motherhood in Western Europe?" Lori B. Andrews expresses the opinion that the new biomedical technologies free and enslave women at the same in an almost invisible manner – using women's own convictions and perceptions. According to Andrews, the risks connected with the growth of the new reproductive technologies and practices are related to unsuitable experiments with women; unstable legal bias, regulating its use; diminishing the meaning of motherhood to "production, in the optimal case, of healthy children"; transforming reproduction to the production process with its own commercial and rational aims.[22]

The pregnancy of a woman is a crucial period, transforming its meaning: from private experience to an object of public control and attention to her decisions and way of life. The womb of the mother transforms into a semi-public territory and the woman's right of autonomy is regarded partly.[23]

In the light of prenatal testing and screening, women have to bear the physical and psychological consequences. In some cases of detected genetic anomalies, the information has a double effect – it reveals the health status not only of the foetus, but also of the mother, and changes not only the perception of the foetus, but also the very self-perception of the mother. Women are more inclined to worry about their test results and to feel guilty in cases of detected mutation. A Swedish survey about the attitudes to prenatal testing reveals that autonomy is more valued by women than by men.[24] 82% of women think that the decision for testing has to be made together with their partners as opposed to 20% of men. 40% of men think that specialists should make the decision. The decision

22 Lorry Andrews, "How is Technology Changing Meaning of Motherhood," in *Women's Reproductive Rights*, ed. Idiakez, I. and Widdows, H. (Palgrave Macmillan, 2006), 124-140.
23 Carol Barnett, "The forgotten and the neglected," *Golden Gate University Law Review* 863 (1993): 886– 887.
24 Berit Sjogren, "Future use and development of prenatal diagnosis, consumers attitudes," *Prenatal Diagnosis* 12 (1992): 1-8.

to refuse testing is more important for women and, unfortunately, is the only form of autonomy, which they can make.

The growing use of prenatal testing is based on the conviction that women have an obligation to guarantee the health of their children. This obligation is often mixed with a sense of guilt:

- When a woman refuses testing and gives birth to a sick child;
- When a woman is obliged to inform her partner about the possibility of defective genes, so to promote his reproductive choice and freedom;
- When a woman undergoes testing and gives birth to a sick child, despite the bad results;
- When a woman undergoes testing and gives birth to a sick child, despite the good results (these are so-called false positive and false negative results).
- When a woman undergoes testing and makes a selective abortion.

Therefore, the very existence and use of such testing influences not only the individual choice, but also the social expectations and perceptions of pregnancy and motherhood. Apart from the fact that genetic disorders are rare, the pregnancy is perceived not as a natural state, but as a territory of risk and pathology. Besides, women are prone to underestimate these risks. This picture reveals the high degree of medicalization and trust in medical technology and authorities in developed countries. The necessity of medicalization is not biological but rather social, so pregnant women are expected to care about what they eat, what they drink, how they sleep, what they feel, how they live by appropriate medical standards for all these activities.[25]

The decision-making process during prenatal genetic testing is a good example of an amalgam between structural (macro) and individual (micro) factors. For medical specialists, the information about the use of prenatal testing is based on professional, legal and ethical norms and guides. From a medical point of view, the important evidence is based on concrete medical facts, statistical information about the detected anomaly and so on. There is a power and informational asymmetry

[25] Rebecca Kukla, *Mass Hysteria: Medicine, Culture, and Mothers' Bodies* (Lanham MD: Rowman and Littlefield., 2005).

between the medics and pregnant women. Pregnant women have to act according to the given information and their emotional balance to cope with it.

The concept of free, individual choice is nevertheless seen as a moral-neutral kind of it. However, the choice is never moral-neutral, especially when it is within the defined limits of someone's life and context. The very process of decision-making on an individual level has its roots in the woman`s reproductive history and visions of future, the so-called imaginary level. This process generates knowledge of the situation. According to the ethical theory, maybe the most appropriate way to conceptualize the practice of prenatal diagnosis is by the methods and framework of ethics of care with its accent on individual narrative and situation.

In the frames of the bioethical discourse, the experience is not only the reason for the emergence of bioethical problems, but also their solution. The moral challenge of the decision-making process on its emotional level has two important faces: relations and emotions. These are the essential accents of the situation as opposed to too normative and abstract ethical theories. This distance leads to a specific kind of ethical conflict, released in three directions, including prenatal testing in national screening programs produces social stigma to concrete disabilities.[26] Pregnant women are so involved in their own situation that they do not realize the morally problematic effects of testing at all, and due to this, the decision-making process is defined rather fragmentally.[27]

Medical specialists with their specific knowledge can see some of the morally problematic tendencies of testing, but their professional place in the structure (health care, maternity care, genetic consultation) creates inner rationality, which complicates their ability to criticize or to change these problematic issues. With its technology and social practice, prenatal diagnosis and testing produce normative effects.[28]

---

[26] Joan Rothschild, *The Dream of the Perfect Child* (Indiana University Press, Bloomington, 2005).

[27] Lucash Kovasc, "Taking Risk in Striving for Certainty. Discrepancies in the Moral Deliberations of Counsellors and Pregnant Women Undergoing PND," in *Ethical Dilemmas in Prenatal Diagnosis*, ed. Tamara Fischmann, Elisabeth Hildt (New York: Springer, 2011), 34.

[28] Hans Harbers and Marcus Popkema, "The Cultural Politics of Prenatal Screening," in *Inside the Politics of Technology: Agency and Normativity in the Co-Production of Technology and Society* (Amsterdam: Amsterdam University Press, 2005), 229-257.

The process of decision-making is the hardest in cases of detected anomalies and a positive diagnosis. At this moment, testing changes its meaning from an instrument of security and control to appeal to a woman's responsibility and consciousness, legitimating the choice of selective abortion. If an offer to test is made, usually during the visit to the maternity care, the pregnant woman cannot make the decision.[29] So the routinization of testing creates specific ethical controversy, insensitive to individual experience and circumstances.

Finally, compared to universal ethical norms, the value of individual narrative can also be under question. The emphasis of individual narrative reduces ethics to the calculation of subjective assessments of the situation and the significance of individual desires. This subjectivity can be morally problematic for a stable bias of moral choices and decisions.

BEYOND MORALITY OR SOME REMARKS ON CHOICE

So the main question of this article remains – is it possible to speak of morality in the context of reproduction, is there a space for real autonomy and integrated and fully understanding of the meaning of pregnancy and birth? Are there any means to empower absolute autonomy for pregnant women and in fact not limit its existence? I consider the role of midwives to be crucial in prenatal and postnatal care, seeing them as guides and supporters for mothers-to-be. I do not believe in the discourse of rights or bans, but I believe in imagination and personal transformation that can bring for the majority of us a sense of common shared values in the understanding of what life is about and. Every one of us, every living being on Earth has the most powerful weapon in their chest – the center of love that reminds us of the meaning of being right here and right now. Our only choice should be not to disturb it, neither by technology nor by the failures of our imagination.

[29] Eva M. Engels, "Experience and Ethics: Ethical and Methodological Reflections on the Integration of the EDIG Study in the Ethical Landscape," in *The Janus Face of Prenatal Diagnostics*, ed. Leuziger-Bohleber M., Engels E.-M., Tsiantis J. (London: Karnac Books, 2008).

CONCLUSION

Individual freedom is a leading principle and moral value in the governance of reproduction due to commercialization and technologization.[30]

On the one side, procreative liberty is not restricted neither by the state, nor by the "important others." The routine use of prenatal testing is evident in the most frequent choices pregnant women make. In the narrative of prenatal testing, norms are the health, the quality of life, the idea of a good life – life without suffering. It is not the disability itself that legitimizes the necessity of testing but the simple fact that anomaly threatens prosperity and the good life. Hence the biological concept of normality is equal to the idea of a good life. The practice of prenatal testing begins to function as a moral genetic code on an individual level – as a desire for normality and on a biopolitical, structural level, as a routinization of concrete, reproductive choices. On the other hand, morality is reduced to the value of genetic health in the frames and restrictions of an individual situation. The current biopolitics of reproduction operates through concepts of risk, insecurity, and individual responsibility, transforming natural states such as pregnancy into a challenge.

REFERENCES

Andrews, L. "How is Technology Changing Meaning of Motherhood." In *Women's Reproductive Rights*, ed. Idiakez, I. and Widdows, H. London: Palgrave Macmillan, 2006.

Apple, R. *Mothers and Medicine: A Social History of Infant Feeding, 1890–1950.* London and Madison: University of Wisconsin Press, 1987.

Aries, P. *Centuries of Childhood.* London: Jonathan Cape, 1962.

Barnett, C. B. "The forgotten and the neglected." *Golden Gate University Law Review* 863 (1993): 886-887.

Berger, P. *Invitation to sociology. A Humanistic Perspective.* London: Anchor, 1999.

Blum, L. *At the Breast: Ideologies of Breastfeeding and Motherhood in the Contemporary United States.* Boston: Beacon Press, 1999.

[30] Catherine Waldby and Melinda Cooper, "The Biopolitics of Reproduction," *Australian Feminist Studies* (2008): 57-73.

Callon, M. and V. Rabeharisoa. "Gino's Lesson on Humanity: Genetics, Mutual Entanglements and the Sociologist's Role." *Economy and Society* 33 (2004): 1-27.

Cole, J. *The Power of Large Numbers: Population, Politics, and Gender in Nineteenth Century France.* Ithaca and London: Cornell University Press, 2000.

De Vries, R., and Benoit C., van Teijlingen E. and Wrede S., eds. *Birth by Design: Pregnancy, Maternity Care, and Midwifery in North America and Europe.* New York: Routledge, 2001.

Deutscher, P. "The Inversion of Exceptionality: Foucault, Agamben, and 'Reproductive Rights'." *South Atlantic Quarterly* 107 (2008): 55-70.

Donzelot, J. *The Policing of Families.* New York: Pantheon, 1979.

Eagleton, T. *The idea of culture.* Hoboken: Wiley, 2000.

Engels, E-M. "Experience and Ethics: Ethical and Methodological Reflections on the Integration of the EDIG Study in the Ethical Landscape." In *The Janus Face of Prenatal Diagnostics*, ed Leuziger-Bohleber M., Engels E.-M., Tsiantis J. London: Karnac, 2008.

Freud, S. *Civilization and its Discontents.* London: Hogarth 1975.

Kovasc, L. "Taking Risk in Striving for Certainty. Discrepancies in the Moral Deliberations of Counsellors and Pregnant Women Undergoing PND." In *Ethical Dilemmas in Prenatal Diagnosis*, ed. Tamara Fischmann, Elisabeth Hildt. New York: Springer, 2011.

Kukla, R. *Mass Hysteria: Medicine, Culture, and Mothers' Bodies.* Lanham MD: Rowman and Littlefield, 2005.

Leuzinger-Bohleber, M., and Engels E-M., Tsiantis J., eds. *The Janus Face of Prenatal Diagnostics – a European study bridging ethics, psychoanalysis and medicine.* London: Karnac Books, 2008.

Litt, J. *Medicalized Motherhood: Perspectives from the Lives of African-American and Jewish Women.* New Brunswick, N.J., and London: Rutgers University Press, 2000.

Lock, M., and P. Kaufert, eds. *Pragmatic Women and Body Politics.* Cambridge: Cambridge University Press, 1998.

Mills, C. "Forthcoming. Resisting Biopolitics, Resisting Freedom: Prenatal Testing and Choice." In *Resisting Biopolitics: Philosophical, Political and Performative Strategies*, ed. Wilmer, Stephen and Zukauskaite, Audrone. London: Routledge, 2015.

Popkema, M., and H. Harbers. *The Cultural Politics of Prenatal Screening. In: Inside the Politics of Technology: Agency and Normativity in the Co-Production of Technology and Society*. Amsterdam University Press, 2005.

Ram, K., and M. Jolly. *Maternities and Modernities: Colonial and Postcolonial Experiences in Asia and the Pacific*. Cambridge: Cambridge University Press, 1998.

Rapp, R. *Testing Women, Testing the Fetus: the Social Impact of Amniocentesis in America*. London: Routledge, 1999.

Rothman, B.K. "Spoiling the pregnancy. Prenatal diagnosis in the Netherlands." In *Birth by design*. London: Routledge, 2001

Rothschild, J. *The Dream of the Perfect Child*. Indiana University Press, Bloomington, 2005.

Sjogren, B. "Future use and development of prenatal diagnosis, consumers attitudes." *Prenatal Diagnosis* 12 (1992): 1-8.

Waldby, C., and M. Cooper. "The Biopolitics of Reproduction." *Australian Feminist Studies* 23(55) (2008): 57-73.

Internet reference link:

"Poland rules abortion due to foetal defects unconstitutional," *The Guardian*. 2020. www.theguardian.com/world/2020/oct/22/poland-rules-abortion-due-to-foetal-defects-unconstitutional?CMP=share_btn_fb&fbclid=IwAR2d_qvzM_iZUM5CN0j8B4H2xf-I1H9QMeZx9YT709Q9ljQGqCPBXu2VrfU.

ETHICAL CHALLENGES IN HUMAN PAPILLOMAVIRUS VACCINATION

Antoanela Naaji,[1] Gratiana Chicin[2]

INTRODUCTION

The Human papillomavirus (HPV) infection is now a global concern. It is the most common sexually transmitted infection that manifests clinically through anogenital warts and it may progress to precancerous lesions or different types of cancers. It is now well established that this infection is the cause of cervical cancer and there is growing evidence of HPV being a relevant factor in other anogenital cancers as well as head and neck cancers.

Data on the HPV role in anogenital cancers other than the cervix are limited, but there are some evidence linking HPV DNA with cancers of the anus, vulva, vagina, and penis. Their association with HPV makes them potentially preventable and subject to similar preventative strategies as those for cervical cancer. According to the HPV Information Centre, About 569,847 new cervical cancer cases are diagnosed annually in the world (estimates for 2018). Cervical cancer ranks as the 3rd leading cause of female cancer and it is the 2nd most common female cancer in women aged 15 to 44 years.[3]

---

[1] Vasile Goldis Western University of Arad, Romania.

[2] National Institute of Public Health, Timisoara, Romania.

[3] Laia Bruni, G. Albero, B. Serrano, M. Mena, D. Gómez, J. Muñoz, Xavier Bosch, Silvia de Sanjosé, *Human Papillomavirus and Related Diseases in the World. Summary Report*. ICO/IARC Information Centre on HPV and Cancer (HPV Information Centre), 17 June 2019. https://www.hpvcentre.net/statistics/reports/XWX.pdf (accessed August, 2019).

Considering that the Human papillomavirus is mostly acquired during adolescence or young adulthood, in order to prevent morbidity and mortality, the World Health Organization (WHO) recommends HPV vaccines as part of routine vaccinations in all countries, along with other prevention measures. Thus, institutionalized vaccination programs have become part of routine health care worldwide.[4]

Monitoring of HPV vaccination coverage is essential in order to assess the performance of vaccination programmes. Since their licensure in 2006, HPV vaccines have been progressively introduced in many countries, mainly targeting girls around the ages of ten to fourteen. Comparison of coverage statistics is limited by differences in age at vaccination, programme delivery strategy, and year. However, global estimates of the extent and impact of vaccine coverage are still unavailable.[5]

According to literature, trials and studies investigated different issues related to HPV vaccination, such as vaccination status and results, the efficiency of human papillomavirus vaccines, parents' HPV vaccine knowledge, attitudes and behavior, ethical issues related to human papillomavirus vaccination programs, HPV vaccination policy, inequalities in cervical cancer burden and prevention, etc. Some of the studies are presented in the "Literature review" section.

In Romania, the Ministry of Health recommends HPV vaccinations based on the fact that countries that included HPV vaccination in their national programmes, reported significant reductions in the prevalence of anogenital warts and different types of cancers. The most important cause for which the HPV types 16 and 18 vaccines was introduced in Romania was the high prevalence of cervical cancer. The current vaccine is for 9 HPV types, including those that cause genital warts. More aspects regarding these issues are presented in "The HPV vaccination in Romania" section.

One of the most important aspects of vaccination in general and more specifically in HPV vaccination is the ethical one. It will be discussed in more detail in the "Ethical Challenges" section.

---

[4] WHO, "Human papillomavirus vaccines: WHO position paper," *WER* 43 (2014), 89, 465–492. www.who.int/wer/2014/wer8943.pdf?ua= (accessed July 2019).

[5] Lauri E. Markowitz, Vivien Tsu, Shelley L. Deeks, et al., "Human papillomavirus vaccine introduction - the first five years," *Vaccine* 30 (suppl. 5) (2010): F139-F148.

LITERATURE REVIEW

According to Bruni et al., worldwide, from June 2006 to October 2014, 64 countries nationally, four countries subnationally, and 12 overseas territories had implemented HPV immunization programs. Between 2006 and 2014, on the global level, about 47 million women received a full course of the HPV vaccine, and 59 million women received at least one dose of the vaccine[6]. For optimum cost-effectiveness, HPV vaccination coverage in women should be about 70%. At a vaccination rate of 50%, a 68% reduction in HPV types 16 and 18 were found and also a 61% decrease in anogenital warts. Thus, 379,000 cases of cervical cancer and 156,000 deaths were prevented until the age of 75. Even if an estimated 118 million women had been targeted through these programmes, only 1% were from low-income or lower-middle-income countries. However, the impact of the vaccine was higher in upper-middle-income countries than in high-income countries, despite the lower number of vaccinated women.[7]

According to the literature, recent trials and studies investigated different issues related to HPV vaccination, such as HPV vaccination policy, vaccination status and results, the efficiency of human papillomavirus vaccines, etc.[8] These studies were carried out all over the

[6] Laia Bruni, G. Albero, B. Serrano, M. Mena, D. Gómez, J. Muñoz, Xavier Bosch, Silvia de Sanjosé, *Human Papillomavirus and Related Diseases in the World. Summary Report.* ICO/IARC Information Centre on HPV and Cancer (HPV Information Centre), 17 June 2019. www.hpvcentre.net/statistics/reports/XWX.pdf (accessed August, 2019).

[7] Laia Bruni, Mireia Diaz, Leslie Barrionuevo-Rosas, Rolando Herrero, Freddie Bray, F. Xavier Bosch, Silvia De Sanjosé, Xavier Castellsagué, "Global estimates of human papillomavirus vaccination coverage by region and income level: a pooled analysis," *Lancet Glob Health* 4(7) (2016): e453-e463.

[8] See the following articles: Catherine De Martel, Martyn Plummer, Jerome Vignat, Silvia Franceschi, "Worldwide burden of cancer attributable to HPV by site, country and HPV type," *International Journal of Cancer* 141(4) (2017): 664-670. Rolando Herrero, Wim Quint, Allan Hildesheim, Paula Gonzalez, Linda Struijk, Hormuzd A Katki, Carolina Porras, Mark Schiffman, Ana Cecilia Rodriguez, Diane Solomon et al. (CVT Vaccine Grp), "Reduced Prevalence of Oral Human Papillomavirus (HPV) 4 Years after Bivalent HPV Vaccination in a Randomized Clinical Trial in Costa Rica," *PLOS ONE* 8(7): 2013. Hammad Ali, Basil Donovan, Handan Wand, Tim R. H. Read, David G. Regan, Andrew E. Grulich, Christopher K. Fairley, Rebecca J. Guy, "Genital warts in young Australians five years into national human papillomavirus vaccination programme: national surveillance data," *British Medical Journal* 346 (2013). Anisa Mburu, Peter Itsura, Hillary Mabeya, Alice Kaaria, Darron R. Brown, "Knowledge of Cervical Cancer and

world. There are also some studies which investigate ethical related issues such as: parents' HPV vaccine knowledge, attitudes and behavior, ethical issues related to human papillomavirus vaccination programs, inequalities in cervical cancer burden and prevention, etc. Furthermore, we will present a few of them.

In "Who calls the shots? The ethics of adolescent self-consent for HPV vaccination,"[9] the authors investigated the factors which influence the rate of human papillomavirus vaccination based on the fact that even the HPV is indicated to reduce the risk of genital warts and certain types of cancer, rates of vaccination are low in the USA. Their hypothesis was that the individual-level factors have an important contribution to low vaccination rates. One of the most important factors was considered to be the need for parental consent. In their study, the authors explored the ethical acceptability of vaccinating adolescents for HPV without parental

---

Acceptability of Prevention Strategies among Human Papillomavirus-Vaccinated and Human Papillomavirus-Unvaccinated Adolescent Women in Eldoret, Kenya," *Bioresearch Open Access* 8(1) (2019): 139-145. Kristina Thorsteinsson, Steen Ladelund, Merete Storgaard, Terese Katzenstein, Isik Somuncu Johansen, Gitte Pedersen, et al., "Persistence of cervical high-risk human papillomavirus in women living with HIV in Denmark - the SHADE," *BMC Infectious Diseases* 19(1) (2019): 740. Tanja Y. Walker, Laurie D. Elam-Evans, James A. Singleton, David Yankey, Lauri E. Markowitz, Benjamin Fredua, Charnetta L. Williams, Sarah A. Meyer, Shannon Stokley, "National, Regional, State, and Selected Local Area Vaccination Coverage Among Adolescents Aged 13-17 Years - United States 2018," *Morbidity and Mortality Weekly Report*, 68(33) (2019): 718-723. Witness Mapanga, Brendan Girdler-Brown, Elvira Singh, "Knowledge, attitudes and practices of young people in Zimbabwe on cervical cancer and HPV, current screening methods and vaccination," *BMC Cancer* 19(1) (2019): 845. Shyamji Rawat, Sulekha Yadav, Pranjil Mandloi, Chinkal Panihar, P.V. Barde, "High Incidence of Human Papillomavirus Types 16 and 18 in Cervical Carcinoma Patients in a Tertiary Care Unit, Jabalpur, MP, India," *Indian Journal of Gynecologic Oncology* 17(3) (2019): 66. Lin Zhu, Shumenghui Zhai, Philip T. Siu, Helen Y. Xia, Sarah Lai, Cristina N. Zambrano, Grace X. Ma, "Factors Related to Chinese Parents' HPV Vaccination Intention for Children," *American Journal of Health Behavior* 43(5) (2019): 994-1005. Bo Zhu, Yunyong Liu, Tingting Zuo, Xiaoli Cui, Mengdan Li, Jing Zhang, Huihuim Yu, Haozhe Piao, "The prevalence, trends, an geographical distribution of human papillomavirus infection in China: The pooled analysis of 1.7 million women," *Cancer Medicine* 8(11) (2019): 5373-5385. Sylvie Abel, Fatiha Najioullah, Jean-Luc Volumenie, Laetitia Accrombessi, Gabriel Carles, Dominique Catherine et al. (HP2V Study Grp), "High prevalence of human papillomavirus infection in HIV-infected women living in French Antilles and French Guiana," *PLOS ONE* 14(9) (2019): e0221334.

[9] Suchi Agrawal, Stephanie R. Morain, "Who calls the shots? The ethics of adolescent self-consent for HPV vaccination," *Journal of Medical Ethics* 44(8) (2018): 531-535.

consent. Based on the analysis that they carried out, the conclusion was that allowing adolescents to self-consent for being immunized by HPV vaccine is justifiable from an ethical point of view and it should be reflected in United States policies.

In "Ethical issues related to human papillomavirus vaccination programs: an example from Bangladesh,"[10] the authors present the current situation in Bangladesh where the HPV vaccine was introduced in 2016, as a pilot project, in Gazipur district, targeting grade five female students and non-school going girls (age range 10-12 years). The paper investigates the various ethical aspects related to the introduction and implementation of such a vaccination program. The study was conducted based on interviews with policy-makers, vaccine implementers, vaccine recipients and ethics specialists from Bangladesh. The authors concluded that more ethical discussion and debate should be done among the public health professionals. Also, adolescent health education should contain more detailed information about HPV, its mode of transmission, risk factors and prevention in order to make an informed choice.

In "Assessing the acceptability of incentivising HPV vaccination consent form return as a means of increasing uptake"[11] the influence of an incentive intervention for obtaining the HPV vaccination consent forms was investigated. The methodology of the study was based on applying a qualitative free-text questionnaire to 181 adolescent girls and 61 parents, assessing the acceptability of an incentive intervention to increase HPV vaccination consent form return. Telephone interviews were also conducted with six members of the staff participating, assessing the acceptability of the incentive. The results of the study showed that the use of an incentive intervention to encourage the return of HPV vaccination consent forms were moderately acceptable. Another conclusion was that improving communication about the nature of the incentive to reduce misconceptions is vital.

[10] Marium Salwa, Tarek Abdullah Al-Munim, "Ethical issues related to human papillomavirus vaccination programs: an example from Bangladesh," *BMC Medical Ethics* 19 (2018).

[11] Lauren Rockliffe, Amanda J. Chorley, Emily McBride, Jo Waller, Alice S. Forster, "Assessing the acceptability of incentivising HPV vaccination consent form return as a means of increasing uptake," *BMC Public Health* 18 (2018).

Another study carried out in Canada[12] examined the 'girls only' policy for publicly funded human papillomavirus (HPV) vaccination programmes, which is criticized by some ethicists as inequitable. This ethical issue is based on some recent studies that indicate increasing rates of other HPV-related cancers among cisgender men and women. The authors consider that these exclusionary practices that leave out those who form their gender identities across the spectrum are unethical. Their conclusion, based on deontological and utilitarian frameworks, is that a collective mobilization is needed in order to fund more inclusive HPV vaccination policies.

The other two studies carried on in 2014 emphasize the role of advertising and professional communication in the decision-making process related to HPV vaccination.

The first study[13] explores the methodological and ethical issues of engaging in health research using social media. As methodology, the authors searched seven electronic databases using search terms related to Human papillomaviruses (HPV) and HPV vaccination. The conclusion was that most researchers did not engage enough virtual users in the research. Most studies did not seek ethical approval from an institutional research board or permission from web service hosts (providers). To respect the principles of concern for welfare, respect for the person, and justice in research using social media, the authors recommend researchers and ethics review boards to work together to develop expertise in evaluating the design of studies where the virtual communities are involved.

The second study[14] was carried out in Sweden and investigated why parents refuse to vaccinate their 10-12 years old daughters with the HPV vaccine. The methodology consisted of interviewing 25 parents. The conclusions showed that, besides other subjective reasons, the main reason was related to the lack or insufficient information necessary for

---

[12] Victoria Law, Diana L. Gustafson, "Challenging 'girls only' publicly funded human papillomavirus vaccination programmes," *Nursing Inquiry* 24(1) (2017).

[13] Diana L. Gustafson, Claire F. Woodworth, "Methodological and ethical issues in research using social media: a meta method of Human Papillomavirus vaccine studies," *BMC Medical Research Methodology* 14 (2014), https://bmcmedresmethodol.biomedcentral.com/articles/10.1186/1471-2288-14-127, (accessed August, 2019).

[14] Maria Grandahl, Marie Oscarsson, Christina Stenhammar, Tryggve Neveus, Ragnar Westerling, Tanja Tyden, "Not the right time: why parents refuse to let their daughters have the human papillomavirus vaccination," *Acta Paediatrica* 103(34) (2014): 436-441.

giving documented informed consent. The results also pointed out that a more flexible HPV vaccination schedule could improve vaccine uptake.

A recent study[15] focuses on the challenges in obtaining the informed consent necessary for carrying out the immunization programs in schools, in England. These programs requested parental informed consent for HPV vaccinations. The methodology of the study consisted of analyzing the results of the data collected from 39 interviews with immunization managers and providers and a descriptive statistical analysis of data resulted from questions related to consent for adolescent vaccination, included in a survey applied to parents and adolescents in 2017. The main conclusions were: (1) parents and adolescents generally agreed on vaccine decisions although only 32% of parents discussed vaccination with their daughters, (2) health professionals were not always clear about the best way to manage the consent process and (3) improving the communication and developing related interventions to improve the logistics and practice of consent in school-based adolescent immunization programmes could help improve uptake.

HPV VACCINATION IN ROMANIA

According to Human Papillomavirus and Related Diseases Report,[16] the HPV infection is the main cause of cervical cancer and, based on the recent evidence, it represents an important factor in other anogenital cancers as well as head and neck cancers. Moreover, HPV types 16 and 18 are responsible for about 70% of all cervical cancer cases globally.

In Romania, about 3,308 new cervial cancer cases are diagnosed annually (estimates for 2018). Cervical cancer ranks as the 2nd leading cause of female cancer in Romania and it is the 2nd most common female cancer in women aged 15 to 44 years. Figure 1 represents a comparison of cervical cancer incidence to other cancers in women of all ages in Romania.

---

[15] Tracey Chantler, Louise Letley, Pauline Paterson, Joanne Yarwood, Vanessa Saliba, Sandra Mounier-Jack, "Optimising informed consent in school-based adolescent vaccination programmes in England: A multiple methods analysis," *Vaccine* 37(36) (2019): 5218-5224.

[16] Laia Bruni, G. Albero, B. Serrano, M. Mena, D. Gómez, J. Muñoz, Xavier Bosch, Silvia de Sanjosé, *Human Papillomavirus and Related Diseases in the World. Summary Report.* ICO/IARC Information Centre on HPV and Cancer (HPV Information Centre), 17 June 2019. www.hpvcentre.net/statistics/reports/XWX.pdf, (accessed August, 2019).

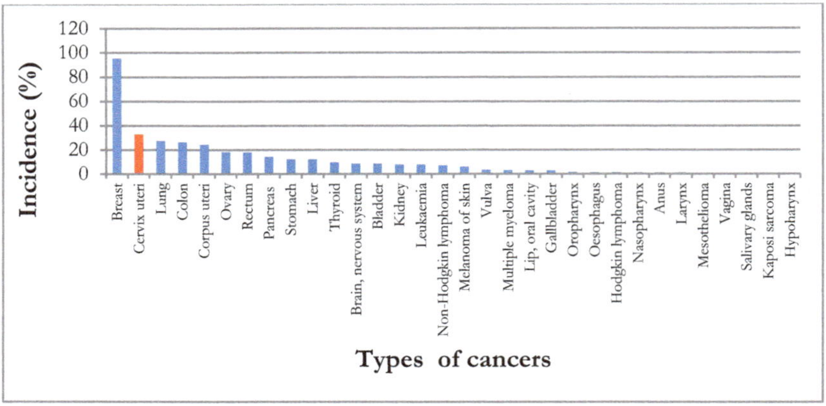

Fig. 1. Incidence of different types of cancers in Romania

In particular, in Romania, cervical cancer represents a critical problem of the public healthcare system, its incidence being very high[17] (Fig. 2).

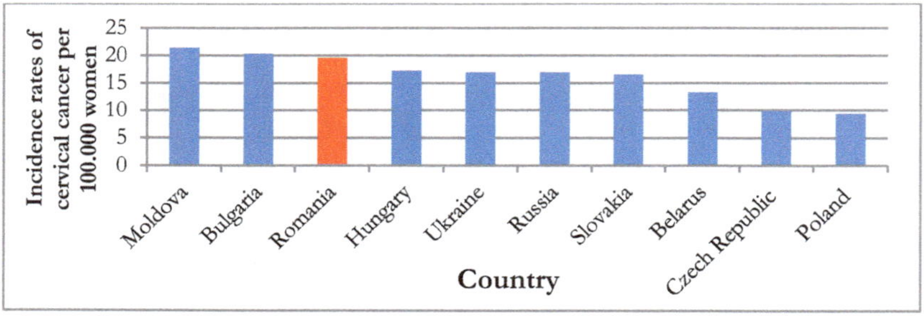

Fig. 2. Age-standardized incidence rates of cervical cancer per 100.000 women

Regarding cervical cancer mortality in Romania, about 1,743 cervical cancer deaths occur annually (estimates for 2018). Cervical cancer ranks as the 3rd leading cause of female cancer deaths in Romania and it is the 2nd leading cause of cancer deaths in women aged 15 to 44 years in Romania.[18] Figures 3 and 4 show a comparison of cervical cancer mortality to other cancers in women of all ages in Romania and a

[17] Ibidem.

[18] Ibidem.

comparison of age-standardized cervical cancer mortality rates in Romania and respectively, in countries within the region.

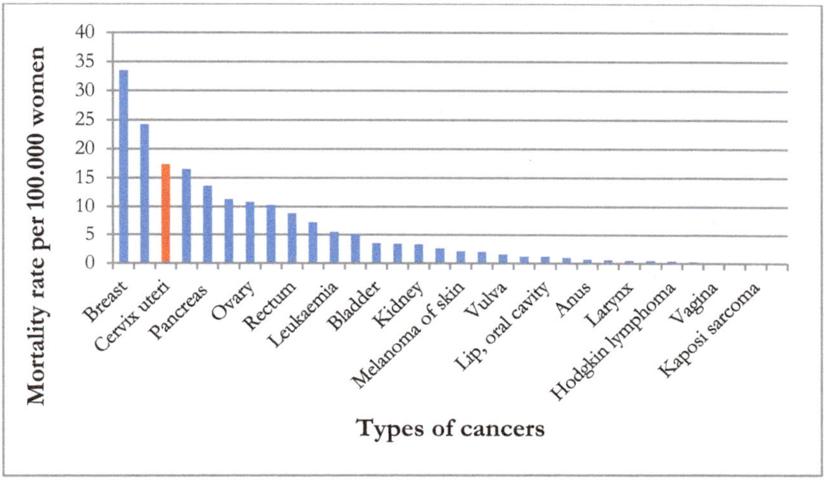

Fig. 3.Mortalityrate caused by different types of cancers per 100.000 women

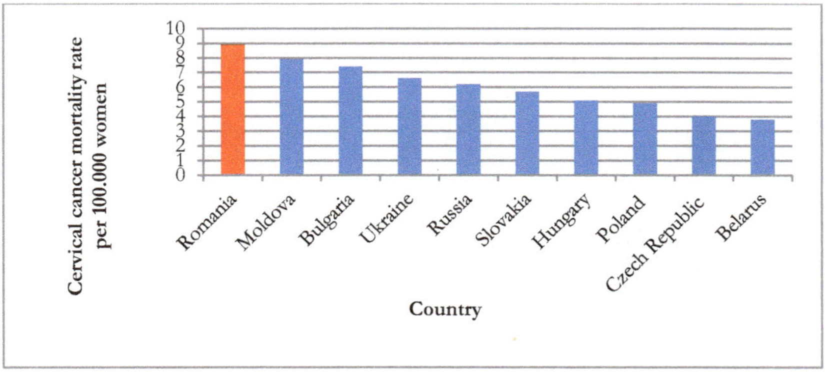

Fig. 4. Age-standardised cervical cancer mortality rates in Romania and countries within the region

The report also presents data related to the influence of HPV in other types of cancers.

Regarding the HPV prevention strategies, it is known that cervical screening programmes can reduce cervical cancer incidence and mortality. Also, there are studies that demonstrated the efficacy of the HPV vaccination in reducing the burden of cervical cancer.

Romania did not have a nationwide, population-based screening programme until 2012 when a National Programme was implemented for women aged 25-64. Therefore, this programme was not a successful one, due to the low rate of attendance and opportunistic participation in screening procedures.[19]

In April 2016, the Romanian Ministry of Health launched the first National plan for control of cancer in Romania, a project that included a comprehensive set of measures for the prevention, reduction of the incidence and mortality of cancer in Romania. Although this plan was in accordance with other programmes existing in Western Europe and it could have a major benefit for cancer patients in Romania, the project has never been implemented.[20]

Regarding the HPV vaccination strategies around the world, in 2006 and 2009 respectively, two human papillomavirus (HPV) vaccines, Gardasil/Silgard (Merck & Co. Inc.) and Cervarix (GlaxoSmithKline, GSK) were licensed by the US Food and Drug Administration. By May 2012, 19 EU countries have implemented HPV vaccination programmes.[21] In 2008, the Romanian Ministry of Health started an immunization campaign in schools, providing free vaccines for 10 to 11

---

[19] Marcela Penta, Adriana Baban, "Mass media coverage of HPV vaccination in Romania: a content analysis," *Health education research* 29(16) (2014): 977-992, www.academia.edu/ 23787954/Mass_media_coverage_of_HPV_vaccination_in_Romania_a_content_analy sis?auto=download, (accessed June, 2019). Iuliana Raluca Gheorghe, "Human papillomavirus (HPV) vaccination campaign in Romania. A case study from a social marketing perspective," in *Advances in Business-Related Scientific Research Conference* (Venice, 2012).

[20] Ramona Gabriela Ursu, Luminita Smaranda Iancu, "Epidemiological Data on HPV Infection in Romania in the Context of the Lack of Correct Cervical Cancer Screening" in *Human Papilloma Virus* (Avid Science, 2017), www.researchgate.net/publication/ 317433912_Epidemiological_Data_on_HPV_Infection_in_Romania_in_the_Context_ of_the_Lack_of_Correct_Cervical_Cancer_Screening, (accessed August, 2019).

[21] *** *Public consultation on draft guidance for introduction of HPV vaccines in EU countries: focus on 9-valent HPV vaccine and vaccination of boys and people living with HIV*, European Centre for Disease Prevention and Control (ECDC) (Stockholm, 2019), www.ecdc.europa.eu/ sites/portal/files/documents/hpv-public-consultation-3-April.pdf, (accessed June, 2019).

years old girls who have received written parental consent. Statistics revealed that only 2.57% of the 110000 girls were vaccinated.

Starting with 2009, an information campaign was launched, followed by a second vaccination programme, targeting 12 to 14 years old girls. Moreover, the programme was made available for adult women, who had the opportunity to get the vaccine free of charge through their health provider. Despite the accessibility of the vaccine, the coverage remained low and the programme for schools was discontinued at the end of 2011. In April 2013, the programme was launched again. At present, HPV vaccination is included in the National Vaccination Program in the category "Vaccination of Population at Risk" and is addressed to girls aged 11–14 years.

Unfortunately, in recent years the vaccination has not been done due to the limited funds and low interest from decision-makers.

ETHICAL CHALLENGES RELATED TO HPV VACCINATION

There are numerous studies that focus on ethical aspects related to HPV vaccination, few of them being presented in the "Literature review" section.

In this section, we present some ethical issues that should be considered in connection with the HPV vaccination and immunization programmes. The main issue is how governments, which are responsible for the health of citizens, and in this case, for immunization programmes, are able to achieve this goal in an equitable and transparent way. One of the most essential means, in order to prevent the spread of HPV infection, is vaccination. From an ethical point of view, the balance between benefits and harms should be established. Also, the autonomy of the individual must be respected, a thing which can lead to a conflict between public and individual interest.[22]

---

[22] *** *Health Technology Assessment (HTA) of HPV vaccination of boys*, Health Information and Quality Authority (Dublin, 2018), www.hiqa.ie/sites/default/files/2018-12/HTA-for-HPV-Vaccination-boys.pdf, (accessed June, 2019).

The principle of autonomy

The principle of autonomy is based on the respect for persons, which means that every person has the right to make his/her own choices. In healthcare systems, the principle of autonomy translates into the principle of informed consent.

In healthcare systems, vaccination is provided to individuals in order to prevent different illnesses, but its benefits may not be visible to the individuals who receive the vaccination, some of them perceiving it as unnecessary. Evaluating the results according to the rate of coverage, the vaccination is sometimes misunderstood. Thus, a high rate of coverage is considered to have a social benefit, and not an individual one. In this case, even in situations where vaccination is not mandated, individuals may feel that they are losing their autonomy. On the other hand, vaccination continues to be a widely accepted public health intervention. Thus, the interplay between autonomy, informed consent and shared decision-making should exist.

The elements of valid informed consent are capacity, understanding of information disclosed and voluntary agreement. In HPV vaccination programmes, one of the most challenging tasks is obtaining consent from adolescents or their parents/guardians. Considering that most of the immunization programmes are targeting children who are too young to have the capacity to consent themselves, parents or guardians are requested to give consent on their behalf. The debate here is if, in order to exercise autonomy, the children should be involved in the decision-making process. Our opinion is that, given the age at vaccination, it may be appropriate to consider the participation of the child in each part of the informed consent process, even if together with the parent. Meanwhile, the informed consent materials must provide sufficient information in a form, manner and language that is understandable to parents. So, in our opinion, the best option for obtaining informed consent for vaccination is to involve both children and parents on the basis of understanding the specific information.

Informed consent is underpinned by the provision of sufficient information. In the absence of clear information then it may be speculated, for example, that the policy decision was on the basis of safety concerns. Also, clear communication in relation to policy decisions on HPV immunization may minimize the risk of misperceptions amongst the public.[23]

Another issue regarding autonomy, in some countries, as mention in a study from Bangladesh,[24] is the use of an implied consent procedure. This means that parents or guardians are informed about vaccination through social media or different official ways of communication, and thus if they send their children to be vaccinated, it means that they are giving the consent to receive the vaccine, and vice versa. An important issue is, even if the consent is implied or written, it should be well informed.

In Romania, the first HPV vaccination campaign has failed due to a lack of communication and information presented in the public space. At the same time, there was no firm position of the authorized institutions about the importance of HPV vaccination, so, in many cases, a misunderstanding of the vaccine's risks/benefits. In this way, the principle of autonomy was respected, but the coverage rate was very low, especially due to the lack of information.

Currently, in Romania, consent is presumed, which means that everyone was supposed to be vaccinated, and only those who refused had to express their choice in writing or verbally. Even so, many of the parents or guardians continued to refuse the vaccine. As a result, the Ministry of Health increased the age of eligible women up to 24 years, then, in 2010, up to 45 years.

The principles of beneficence and non-maleficence

Beneficence requires that the procedure be provided in the patient's interest, with the intent of doing good. Thus, the healthcare providers should develop and maintain skills and knowledge, continually update

---

23 *** *Health Technology Assessment (HTA) of HPV vaccination of boys*, Health Information and Quality Authority (Dublin, 2018), www.hiqa.ie/sites/default/files/2018-12/HTA-for-HPV-Vaccination-boys.pdf, (accessed June, 2019). Jeroen Luyten, Bart Engelen, Philippe Beutels, "The Sexual Ethics of HPV Vaccination for Boys," in *HEC Forum* (Springer, 2013), https://pure.uvt.nl/ws/portalfiles/portal/5039613/luyten_engelen_beutels_sexual_ethics_of_hpv_vaccination_for_boys.pdf, (accessed June, 2019).

24 Marium Salwa, Tarek Abdullah Al-Munim, "Ethical issues related to human papillomavirus vaccination programs: an example from Bangladesh," *BMC Medical Ethics* 19(1) (2018).

training, consider individual circumstances of all patients, and strive for net benefit.

Non-maleficence requires that a procedure does not harm the patient involved or others in society. In some cases, this principle is difficult for doctors to fully apply it due to some conditions or factors which are influencing the medical act. In practice, sometimes, "the do no harm principle" is understood as minimizing as much as possible the harm and maximize the benefits.

In the case of HPV vaccination, the benefit-harm balance must be considered not only at the population level, translated into coverage rate, but also at the individual level. The decision to be vaccinated is made by individuals, based on the evaluation of risks and benefits, perceived personally. Thus, the policy decision-maker should consider this aspect even if they consider the benefit-harm balance at the population level.

Based on the existing literature, the only harms that could be generated by vaccination are related adverse events, which usually are non-serious reactions, such as soreness or redness at the injection site. Only a small minority reported severe adverse events, such as anaphylactic reactions.

One of the difficulties in evaluating the harm is due to collecting clinical outcome data from children and adolescents who receive the vaccine. However, non-serious adverse events can be noticed after the vaccination, so the quantification of such harm is easy to asses.

Regarding the benefits, an important issue is that the time lag between vaccination and future benefits could be very long. In the case of HPV-related tumors, onset is typically decades after vaccination, so the effects of the implementation of an HPV immunization programme could take many years before the impact is observed. This means that it may never be possible to accurately determine the full benefits of vaccination.

Therefore, in countries where HPV vaccination was implemented, there is international evidence that for anogenital warts, the benefits are more evident through the decline of the incidence of these infections. Considering other prevention strategies, like cervical screening programmes, which are in the real benefit of the targeted population by reducing morbidity and mortality, both strategies should be applied.

In Romania, when the first HPV immunization programme was introduced, there were opinions against the HPV vaccination. The main arguments of adolescents or their parents who refused to enter HPV immunization programmes were: the insufficient tests and fear of adverse effects. Also, an assumption that the entire program was an experiment of big pharmaceutical companies in the local population circulated during that period. Thus, in the first year, only 2.5% of the families of the targeted population accepted the vaccination, and in 2010-2011 the Romanian state began to destroy the purchased doses of vaccine, because they expired. All these facts have had a negative impact on the correct understanding of the benefits of the HPV vaccination[25].

As conclusion, vaccination is often used as a mechanism to achieve benefits for the greater good, even if only a few will experience a substantial benefit to the detriment of many individuals who experience a minor burden. In the case of HPV vaccination, the incidence of HPV-attributable diseases being higher, a larger proportion of the population will experience some form of benefits.

The principle of justice

The principle of justice requires, in the medical field, a fair distribution of services. The health care provider must consider when evaluating justice some issues such as fair distribution of resources, competing needs, rights and obligations, and potential conflicts with the established legislation.

Regarding this principle, there is an ethical dilemma because in many countries the HPV vaccination policies are available only for girls, so the immunization programmes are not equally available to all the targeted people. Considering the fact that boys can also benefit from vaccination, the existing girls-only programmes may be viewed as discriminating against boys. When HPV vaccination was initially introduced, it was as a means to reduce the incidence of cervical cancer and hence seen as benefiting girls only. The HPV vaccination of boys is a very important issue that could be analyzed in a future study.

The evaluation of how the principle of justice is applied in the case of HPV immunization programmes, should be done according to the financial resources of each country.

---

[25] *** *Public consultation on draft guidance for introduction of HPV vaccines in EU countries: focus on 9-valent HPV vaccine and vaccination of boys and people living with HIV*, European Centre for Disease Prevention and Control (ECDC) (Stockholm, 2019), www.ecdc.europa.eu/sites/portal/files/documents/hpv-public-consultation-3-April.pdf, (accessed June, 2019).

Other ethical considerations

One of the ethical aspects which are less discussed in literature is related to the use of an intervention to encourage the HPV vaccination consent. In one of the mentioned studies[26], the conclusion was that incentives were moderately acceptable to those receiving and delivering the intervention and according incentives could be an acceptable means of improving HPV vaccination rates.

Another ethical issue, which we consider important in case of HPV vaccination, is respecting the privacy of individuals, which of course, should be part of the informed consent. The non-discrimination restriction will direct selective vaccination policies as much as possible towards relevant characteristics at the individual level. However, this can lead to infringements of privacy, since characteristics like sexual activity or other behaviours, in which disease transmission can occur, typically belong to an individual's privacy.

CONCLUSIONS

As conclusion, we consider that HPV vaccination is an essential measure in preventing anogenital warts and different types of cancers, but it must be done in compliance with ethical principles.

An important issue from an ethical point of view, in case of HPV immunization programmes, is obtaining informed consent. Considering that the target population is usually underage, the information of the parents/guardians must be done correctly. Thus, medical authorities should organize information campaigns that encourage vaccination and, at the same time, annihilate campaigns that misinform, especially if they are not based on studies and are not scientifically substantiated.

For a successful HPV immunization programme and for obtaining a high HPV vaccination coverage, communication remains one of the most important aspects of the advertising campaigns.

---

[26] Lauren Rockliffe, Amanda J. Chorley, Emily McBride, Jo Waller, Alice S. Forster, "Assessing the acceptability of incentivising HPV vaccination consent form return as a means of increasing uptake,"*BMC Public Health* 18 (2018).

REFERENCES

Abel, Sylvie, and Najioullah Fatiha, Volumenie Jean-Luc, Accrombessi Laetitia, Carles Gabriel, Catherine Dominique et al. (HP2V Study Grp). "High prevalence of human papillomavirus infection in HIV-infected women living in French Antilles and French Guiana." *PLOS ONE* 14:9 (2019).

Agrawal, Suchi, and Morain Stephanie R. "Who calls the shots? The ethics of adolescent self-consent for HPV vaccination." *Journal of Medical Ethics* 44:8 (2018): 531-535.

Ali, Hammad, and Donovan Basil, Wand Handan, Read Tim R. H., Regan David G., Grulich Andrew E., Fairley Christopher K., Guy Rebecca J. "Genital warts in young Australians five years into national human papillomavirus vaccination programme: national surveillance data." *British Medical Journal* 346 (2013).

Bruni, Laia, and Albero G., Serrano B., Mena M., Gómez D., Muñoz J., Bosch X., de Sanjosé Silvia. *Human Papilloma virus and Related Diseases in the World. Summary Report.* ICO/IARC Information Centre on HPV and Cancer (HPV Information Centre), 17 June 2019. www.hpv centre.net/statistics/reports/XWX.pdf. Accessed August, 2019.

Bruni, L., and Albero G., Serrano B., Mena M., Gómez D., Muñoz J., Bosch F.X., de Sanjosé S. *Human Papillomavirus and Related Diseases in Romania. Summary Report 17 June 2019.* ICO/IARC Information Centre on HPV and Cancer (HPV Information Centre), https://hpv centre.net/statistics/reports/ROU.pdf (accessed August, 2019).

Bruni, Laia, and Diaz Mireia, Barrionuevo-Rosas Leslie, Herrero Rolando, Bray Freddie, Bosch Xavier, De Sanjosé Silvia, Castellsagué Xavier. "Global estimates of human papillomavirus vaccination coverage by region and income level: a pooled analysis." *Lancet Glob Health* 4:7 (2016): e453-e463.

Chantler, Tracey, and Letley Louise, Paterson Pauline, Yarwood Joanne, Saliba Vanessa, Mounier-Jack Sandra. "Optimising informed consent in school-based adolescent vaccination programmes in England: A multiple methods analysis." *Vaccine* 37:36 (2019): 5218-5224.

De Martel, Catherine, and Plummer Martyn, Vignat Jerome, Franceschi Silvia. "Worldwide burden of cancer attributable to HPV by site, country and HPV type." *International Journal of Cancer* 141:4 (2017): 664-670.

Gheorghe, Iuliana Raluca. "Human papilloma virus (HPV) vaccination campaign in Romania. A case study from a social marketing perspective." In *Advances in Business-Related Scientific Research Conference.* Venice, 2012.

Gustafson, Diana L., and Woodworth Claire F. "Methodological and ethical issues in research using social media: a metamethod of Human Papillomavirus vaccine studies." *BMC Medical Research Methodology* 14 (2014), https://bmcmedresmethodol.biomedcentral.com/articles/10.1186/1471-2288-14-127.

Grandahl, Maria, and Oscarsson Marie', Stenhammar Christina, Neveus Tryggve, Westerling Ragnar, Tyden Tanja. "Not the right time: why parents refuse to let their daughters have the human papillomavirus vaccination." *Acta Paediatrica* 103:4 (2014): 436-441.

Herrero, Rolando, and Quint Wim, Hildesheim Allan, Gonzalez Paula, Struijk Linda, Katki Hormuzd A., Porras Carolina, Schiffman Mark, Cecilia Rodriguez Ana, Solomon Diane et al. (CVT Vaccine Grp). "Reduced Prevalence of Oral Human Papillomavirus (HPV) 4 Years after Bivalent HPV Vaccination in a Randomized Clinical Trial in Costa Rica." *PLOS ONE* 8:7 (2013).

Law, Victoria G., and Gustafson Diana L. "Challenging 'girls only' publicly funded human papillomavirus vaccination programmes." *Nursing Inquiry* 24:1 (2017).

Luyten, J., and Engelen B., Beutels P. "The Sexual Ethics of HPV Vaccination for Boys." In *HEC Forum.* Springer, 2013. https://pure.uvt.nl/ws/portalfiles/portal/5039613/luyten_engelen_beutels_sexual_ethics_of_hpv_vaccination_for_boys.pdf. Accessed June, 2019.

Mapanga, Witness, and Girdler-Brown Brendan, Singh Elvira. "Knowledge, attitudes and practices of young people in Zimbabwe on cervical cancer and HPV, current screening methods and vaccination." *BMC Cancer* 19:1 (2019).

Markowitz, L.E., and Tsu V., Deeks S.L. et al. "Human papillomavirus vaccine introduction - the first five years." *Vaccine* 30 (suppl. 5) (2012): F139-F148.

Mburu, Anisa, and Itsura Peter, Mabeya Hillary, Kaaria Alice, Brown Darron R. "Knowledge of Cervical Cancer and Acceptability of Prevention Strategies Among Human Papillomavirus-Vaccinated and

Human Papillomavirus-Unvaccinated Adolescent Women in Eldoret, Kenya." *Bioresearch Open Access* 8:1 (2019): 139-145.

Penta, Marcela, and Baban Adriana. "Mass media coverage of HPV vaccination in Romania: a content analysis." *Health education research* 29:16 (2014): 977-992.

Rawat, Shyamji, and Yadav Sulekha, Mandloi Pranjil, Panihar Chinkal, Barde P.V. "High Incidence of Human Papillomavirus Types 16 and 18 in Cervical Carcinoma Patients in a Tertiary Care Unit, Jabalpur, MP, India." *Indian Journal of Gynecologic Oncology* 17:3 (2019).

Rockliffe, Lauren, and Chorley Amanda J., McBride Emily, Waller Jo, Forster Alice S. "Assessing the acceptability of incentivising HPV vaccination consent form return as a means of increasing uptake." *BMC Public Health* 18:382 (2018).

Salwa, Marium, and Al-Munim Tarek Abdullah. "Ethical issues related to human papillomavirus vaccination programs: an example from Bangladesh." *BMC Medical Ethics* 19:1 (2018).

Thorsteinsson, Kristina, and Ladelund Steen, Storgaard Merete, Katzenstein Terese, Johansen Lsik Somuncu, Pedersen Gitte et al. "Persistence of cervical high-risk human papillomavirus in women living with HIV in Denmark - the SHADE." *BMC Infectious Diseases* 19:1 (2019).

Ursu, Ramona Gabriela, and Iancu Luminita Smaranda. "Epidemiological Data on HPV Infection in Romania in the Context of the Lack of Correct Cervical Cancer Screening." In *Human Papilloma Virus*. Avid Science, 2017. www.researchgate.net/publication/317433912_Epidemiological_Data_on_HPV_Infection_in_Romania_in_the_Context_of_the_Lack_of_Correct_Cervical_Cancer_Screening. Accessed August, 2019.

Walker, Tanja Y., and Elam-Evans Laurie D., Singleton James A., Yankey David, Markowitz Lauri E., Fredua Benjamin, Williams Charnetta L., Meyer Sarah A., Stokley Shannon. "National, Regional, State, and Selected Local Area Vaccination Coverage Among Adolescents Aged 13-17 Years - United States 2018." *Morbidity and Mortality Weekly Report* 68:33 (2019): 718-723.

Zhu, Lin, and Zhai Shumenghui, Siu Philip T., Xia Helen Y., Lai Sarah, Zambrano Cristina N., Ma Grace X. "Factors Related to Chinese Parents' HPV Vaccination Intention for Children." *American Journal of Health Behavior* 43:5 (2019): 994-1005.

Zhu, Bo, and Liu Yunyong, Zuo Tingting, Cui Xiaoli, Li Mengdan, Zhang Jing, Yu Huihuim, Piao Haozhe. "The prevalence, trends, and geographical distribution of human papillomavirus infection in China: The pooled analysis of 1.7 million women." *Cancer Medicine* 8:11 (2019): 5373-5385.

ETHICAL ISSUES AND SPIRITUAL CHALLENGES IN END-OF-LIFE CARE

Danijela Vučević,[1] Igor Pantić,[2] Janko Samardžić,[3] Bojan Jorgačević,[1] Jovana Paunović Pantić,[1] Tatjana Radosavljević[1]

INTRODUCTION

All humans are mortals and death is an inevitable occurrence.[4] Dying is not primarily a biomedical event.[5] It is an intimate personal experience with profound psychological, interpersonal and spiritual meanings.[6] From the biomedical perspective, the end of life may be defined as that time when death, whether due to terminal illness, or acute or chronic illness, is expected within hours to months and can no longer be

---

[1] Institute of Pathophysiology "Ljubodrag Buba Mihailović," Faculty of Medicine, University of Belgrade, Serbia.

[2] Laboratory for Cellular Physiology, Institute of Medical Physiology "Rihard Burian," Faculty of Medicine, University of Belgrade, Serbia. University of Haifa, Israel

[3] Institute of Pharmacology, Clinical Pharmacology and Toxicology, Faculty of Medicine, University of Belgrade, Serbia.

[4] Sameera Karnik and Amar Kanekar, "Ethical issues surrounding end-of-life care: a narrative review," *Healthcare (Basel)* 4 (2016): 24 doi:10.3390/healthcare4020024

[5] Michael Rabow, Steven Pantilat, Scott Steiger, and Ramana Naidu, "Palliative care and pain management," in *Current Medical Diagnosis and Treatment*, eds. Maxine Papadakis and Stephen McPhee (New York: Mc Graw Hill, 2018), 72-99.

[6] Stephen Scher and Kasia Kozlowska, *Rethinking health care ethics eBook*, doi: 10.1007/981-13-0830-7 (New York: Palgrave Pivot, 2018). Holly Nelson-Becker, Amy Ai, Faith Hopp, Thomas McCormick, Judith Schlueter, and Jessica Camp, "Spirituality and religion in end-of-life care ethics: the challenge of interfaith and cross-generational matters," *The British Journal of Social Work* 45 (2015): 104-119. Lindsay Carey and Timothy Hodgson, "Chaplaincy, spiritual care and moral injury: Considerations regarding screening and treatment," *Frontiers in Psychiatry* 9 (2018): doi: 10.3389/fpsut.2018.00619.

reasonably forestalled by medical intervention.[7] Advancements in medical technology are changing the norms of natural death.[8] These technologically advanced treatments have the capability to intervene at the time of death and prolong the lives of people. Biomedical technologies are facilitating to reshape the circumstances around natural death, by sustaining human lives.[9] Although medical treatments have advanced technologically, they hold no promises for recovery.[10] Moreover, they can sustain life with or without meaningful existence, or with secondary support (like feeding tubes, ventilators, etc.).[11] Hence, these medical advancements have empowered patients and their families with the important task of choosing their treatment preference during end-of-life care. In such cases, palliative care may become the sole focus of care, since its main goal is to improve patients' quality of life.[12]

On a global scale, more than 20 million persons are in need of palliative care at the end of life annually.[13] Additionally, 18 million die in unnecessary pain and distress, also impacting the health and well-being of their family members and caretakers.[14] Palliative care, as interdisciplinary care (medicine, nursing, social work, chaplaincy, and other specialties when appropriate) includes management of physical symptoms (pain, dyspnea, nausea, vomiting, constipation and agitation), emotional distress (depression, anxiety and interpersonal strain), and existential distress, such as spiritual crisis.[15] Simultaneously, palliative care supports patients' families and loved ones and helps align patients' care with their preferences and goals. Ideally, this care is initiated at the

---

[7] Michael Rabow, et al., "Palliative care and pain management," 72-99.

[8] Ibidem.

[9] Randall Curtis and Jean-Louis Vincent, "Ethics and end-of-life care for adults in the intensive care unit," *Lancet* 376 (2010): 1347-1353.

[10] Sameera Karnik and Amar Kanekar, "Ethical issues surrounding end-of-life care," 24.

[11] Ibidem. Randall Curtis and Jean-Louis Vincent, "Ethics and end-of-life care for adults in the intensive care unit," 1347-1353.

[12] Bidhu Mohanti, "Ethics in palliative care," *Indian Journal Palliative Care* 15 (2009): 89-92.

[13] Amy Kelley and Sean Morrison, "Palliative care for seriously ill," *The New England Journal of Medicine* 373 (2015): 747-755.

[14] Ellen Fink-Samnick, "The evolution of end-of-life care. Ethical implications for case management," *Professional Case Management* 21 (2016): 180-192.

[15] Michael Rabow, et al., "Palliative care and pain management," 72-99. Bidhu Mohanti, "Ethics in palliative care," 89-92. Amy Kelley and Sean Morrison, "Palliative care for seriously ill," 747-755.

time of diagnosis and is provided concordantly with all other disease-directed or curative treatments.[16]

AGING "TSUNAMI" AS A GLOBAL PROBLEM

> My older patients are like rich paintings, and do they have stories to tell. On my best days I can look at them and see all the way back to their childhood. I think of their parents (long gone now), the places they've been, the things they've seen. To me, it's like looking through the other end of telescope, back to the beginning.
>
> David Dosa, geriatrician[17]

Caring for the aging population is one of the most important challenges of the twenty-first century.[18] Media outlets, political figures, books, and television programming frequently elaborate on the difficulties the world will face as the age of the population increases.[19] Studies also indicate that not only are people living to an older age, but they may also have an increased number of quality years, as defined by better cognitive function and self-perceived health at the end of life.[20]

According to the World Health Organization (WHO), 71 years was the average life expectancy of the global population in 2013.[21] However, the average life expectancy in developing countries is as low as 50 years.[22] Most developed countries use the age of 65 years as the cutoff to define

---

[16] Amy Kelley and Sean Morrison, "Palliative care for seriously ill," 747-755. Michael Rabow, et al., "Palliative care and pain management," 72-99.

[17] David Dosa, *Making Rounds with Oscar: The Extraordinary Gift of an Ordinary Cat* (New York: Hyperion, 2010).

[18] Kristin Oliveira and Linda Maerz, "Epidemiology of critical illness in the elderly," in *Principles of Geriatric Critical Care,* eds. Shamsuddin Akhtar and Stanley Rosenbaum (Cambridge, United Kingdom: Cambridge University Press, 2018), 1-10.

[19] Michael Gusmano and Sara Allin, "Framing the issue of ageing and health care spending in Canada, the United Kingdom and the United States," *Health Economics, Policy and Law* 9 (2014): 313-328.

[20] Kristin Oliveira and Linda Maerz, "Epidemiology of critical illness in the elderly," 1-10. Michael Gusmano and Sara Allin, "Framing the issue of ageing and health care spending in Canada, the United Kingdom and the United States," 313-328.

[21] Kristin Oliveira and Linda Maerz, "Epidemiology of critical illness in the elderly," 1-10.

[22] World Health Organization, *Global health observatory data: life expectancy*, 2013, available at www.who.int/gho/mortality_burden_disease/life_tables/situation_trends_text/en/ (accessed February 13, 2021).

an older person, but in developing countries this definition does not necessarily correlate.[23]

Aging is a chronologic biologic certainty, but it is also subject to the social constructs of cultures.[24] In developed countries, such as the United States, chronologic age plays a paramount role.[25] However, in developing countries, the change in social roles plays a far more important part in defining the elderly.[26] In most developed countries, 60 or 65 years is the age of retirement and is therefore often the cutoff for the definition of old age.[27] In contrast, in developing countries, where retirement is less common, social roles and loss of social station due to physical decline play a far more critical part in dictating the definition of old age.[28] Therefore, in developing countries, old age commences at the stage of life where one is no longer able to contribute to one`s assigned role in society.[29]

According to the UN *World Population Prospects,* in 2015, the world population reached 7.3 billion, an increase of nearly 1 billion over the last 12 years.[30] It is estimated that the world population is growing at an average rate of 1.18 percent per year.[31] Population growth has been particularly high in the least developed countries.[32] Continents with high fertility rates, such as Africa, have had the highest rate of population growth in the last decade (2.5 percent annually).[33]

---

[23] Kristin Oliveira and Linda Maerz, "Epidemiology of critical illness in the elderly," 1-10.

[24] Ibidem.

[25] Anthony Glascock and Susan Feinman, "A holocultural analysis of old age," *Comparative Social Research* 3 (1980): 311-332.

[26] Anthony Glascock and Susan Feinman, "A holocultural analysis of old age," 311-332.

[27] Kristin Oliveira and Linda Maerz, "Epidemiology of critical illness in the elderly," 1-10.

[28] Ebrahim Shah, "The ageing and development report 1999: poverty, independence and the world`s older people," *British Medical Journal* 321 (2000): 517.

[29] Ibidem.

[30] United Nations Department of Economic and Social Affairs, *World population prospects: the 2015 revision*, 2015, available at http://esa.un.org/unpd/wpp/ (accessed February 13, 2021).

[31] American Statistical Association, *World population likely to surpass 11 billion in 2100: US population projected to grow by 40 percent over next 85 years, 2015*, available at www.sciencedaily.com/releases/2015/08/150810110634.htm (accessed February 13, 2021).

[32] Kristin Oliveira and Linda Maerz, "Epidemiology of critical illness in the elderly," 1-10.

[33] United Nations Department of Economic and Social Affairs, *World population prospects: the 2015 revision*, 2015, available at http://esa.un.org/unpd/wpp/ (accessed February 13, 2021).

Given the increased number of industrialized countries in the world, the population has aged at unprecedented rates. Predictably, over the last ten years, life expectancy has increased globally by 3 years (from 67 to 70 years).[34] The most significant increase in life expectancy occurred in Africa. However, this increase is still significantly lower than the North American life expectancy (60 versus 79 years).[35] As a result of increased life expectancy, the population group older than 60 years of age has had the largest annual growth rate at a global level (3.2 percent per year), nearly three times that of the population as a whole.[36] In 2015, the United Nations reported that globally, there were 901 million people older than 60 years of age, which comprised nearly 12 percent of the world population.[37]

Elderly patients, especially those who are more than 80 years old, are one of the fastest-growing segments of the population.[38] Population demographics are changing, and the proportion of elderly patients who are being treated for critical illness continues to increase rapidly.[39] The United Nations estimates that by 2050, the number of elderly in the world will, for the first time in history, be greater than the number of young individuals.[40] Increasing numbers of elderly patients are undergoing surgical procedures that decades ago would have been considered prohibitively high risk and would not be offered to elderly patients.[41] Thus, many elderly patients require intensive monitoring and postoperative care in critical care units.

---

[34] United Nations Department of Economic and Social Affairs, *World population prospects: the 2015 revision*, 2015, available at http://esa.un.org/unpd/wpp/ (accessed February 13, 2021). Kristin Oliveira, Linda Maerz, "Epidemiology of critical illness in the elderly," 1-10.

[35] Ibidem.

[36] United Nations Department of Economic and Social Affairs, *World population prospects: the 2015 revision*, 2015, available at http://esa.un.org/unpd/wpp/ (accessed February 13, 2021).

[37] United Nations Department of Economic and Social Affairs, *World population prospects: the 2015 revision*, 2015, available at http://esa.un.org/unpd/wpp/ (accessed February 13, 2021). Kristin Oliveira and Linda Maerz, "Epidemiology of critical illness in the elderly," 1-10.

[38] Shamsuddin Akhtar and Stanley Rosenbaum, "Preface," in *Principles of Geriatric Critical Care*, eds. Shamsuddin Akhtar and Stanley Rosenbaum (Cambridge, United Kingdom: Cambridge University Press, 2018).

[39] Shamsuddin Akhtar and Stanley Rosenbaum, "Preface."

[40] Kristin Oliveira and Linda Maerz, "Epidemiology of critical illness in the elderly," 1-10.

[41] Shamsuddin Akhtar and Stanley Rosenbaum, "Preface."

Geriatric patients develop significant physiologic changes with aging, including an increased incidence of frailty and diminished physiologic reserve.[42] Complicating the care of geriatric patients are concurrent multiple comorbidities and a high rate of polypharmacy.[43] Elderly patients are more likely to develop chronic critical illness and cognitive dysfunction and to consume significant healthcare resources. Recovery from critical illness implies not only physical recovery and survival, but also a successful return to baseline function and quality of life. Unfortunately, many elderly patients are unable to achieve these goals after acute illness.[44]

ETHICAL ISSUES AT THE END OF LIFE

The clinicians' care of patients at the end of life is guided by the same ethical principles that inform other types of medical care.[45] Foremost among these is truthtelling, nonmaleficence (to not harm), beneficence (to do good), autonomy (to respect individuals' rights to make their own decisions), confidentiality (to follow through and to keep promises), and procedural and distributive justice (to treat other fairly). In order to promote patient autonomy, clinicians are obligated to inform patients about risks, benefits, alternatives and expected outcomes of medical interventions, such as cardiopulmonary resuscitation (CPR), mechanical ventilation, hospitalization and treatment in an intensive care unit (ICU), as well as artificial nutrition and hydration.[46]

Physicians are all too often under the public spotlight for the ethical dilemmas they face in end-of-life care.[47] For example, many treatments that promote beneficence and autonomy, such as surgery or bone marrow transplantation, may violate the clinician's obligation for nonmaleficence.[48] Similarly, while patients may express their autonomy

[42] Ibidem.

[43] Sameera Karnik and Amar Kanekar, "Ethical issues surrounding end-of-life care: a narrative review," 24. Michael Rabow, et al., "Palliative care and pain management," 72-99.

[44] Shamsuddin Akhtar and Stanley Rosenbaum, "Preface."

[45] Michael Rabow, et al., "Palliative care and pain management," 72-99.

[46] Michael Rabow, et al., "Palliative care and pain management," 72-99. Randall Curtis and Jean-Louis Vincent, "Ethics and end-of-life care for adults in the intensive care unit," 1347-1353.

[47] Michael Rabow, et al., "Palliative care and pain management," 72-99.

[48] Ibidem.

as a desire for a particular medical intervention, such as CPR in the setting of multisystem organ failure, the clinicians may decline to provide the intervention since it is of no therapeutic benefit and thus violates beneficence and nonmaleficence.[49] On the other hand, because such unilateral actions violate the autonomy of patients, clinicians should rarely resort to do such a thing. However, clinicians must use caution in invoking futility since strict futility is rare and what constitutes futility is often a matter of controversy.[50]

Futility is often raised as a justification for withdrawing and withholding decisions.[51] Treatment that is perceived to be futile is common in the ICU setting.[52] In one study, approximately 20 percent of patients in the ICU were adjudged to have received futile (11 percent), or probably futile (8.6 percent) care.[53]

There is little if any disagreement among ethicists or clinicians that truly futile therapy need not be offered, should not be knowingly undertaken, and is probably actually unethical.[54] Care that fails to meet a patient's goals, or that maintains them in a suspended state of intensive therapy, but the minimal function is not only costly, but also runs counter to professional values in medicine and creates serious moral conflicts for all concerned. Finding consensus regarding exactly what futility is, however has proven problematic.[55]

When a therapy presents a patient with more burdens than benefits, it should trigger a discussion about the goals of therapy.[56] Defining a treatment as futile implies that there are few or no benefits to consider (or that at least the primary goal of the therapy will not be met) and therefore the treatment will result primarily in harms. Offering such

---

49 Michael Rabow, et al., "Palliative care and pain management," 72-99. Randall Curtis and Jean-Louis Vincent, "Ethics and end-of-life care for adults in the intensive care unit," 1347-1353.

50 Gail Van Norman, "Ethical issues: withdrawing, withholding, and futility," in *Principles of Geriatric Critical Care*, eds. Shamsuddin Akhtar and Stanley Rosenbaum (Cambridge, United Kingdom: Cambridge University Press, 2018), 200-209.

51 Gail Van Norman, "Ethical issues: withdrawing, withholding, and futility," 200-209.

52 Ibidem.

53 Thanh Huynh, Eric Kleerup, Joshua Wiley, Terrance Savitsky, Diana Guse, Bryan Garber, and Neil Wenger, "The frequency and cost of treatment perceived to be futile in critical care," *JAMA Internal Medicine* 173 (2013): 1887-1894.

54 Gail Van Norman, "Ethical issues: withdrawing, withholding, and futility," 200-209.

55 Ibidem.

56 Ibidem.

treatment would be unethical because it would violate principles of beneficence and nonmaleficence, both of which are pillar concepts in Western medical ethics. Intentionally providing futile therapy furthermore threatens the value of veracity in medical practice.[57] Moreover, offering or pursuing futile treatment encourages false hopes.[58] Besides, the concept of futility may have limited usefulness at the bedside, but is a critical concept in understanding the perspectives of multiples parties in withdraw or withhold decisions.[59]

The ICU presents daily opportunities for ethical dilemmas and discussions. Issues run the gamut from competency and informed consent to surrogate decision making, futility, withdrawing and withholding of life-sustaining treatments (LSTs) and vital organ donation.[60]

Withholding an LST refers to a decision to not start or escalate an intervention, and withdrawing an LST refers to discontinuing therapy that has already begun.[61] In the ICU, the withdrawal and withholding of decisions concern a wide spectrum of interventions, e.g., artificial hydration and nutrition, ventilator therapy, CPR, and pacemaker or implantable cardioverter-defibrillator therapies.[62] Regardless of the therapy involved, ethical considerations underlying a withdrawing or withholding of a decision stand on similar moral reasoning. Despite long-standing consensus, there is still considerable disquiet among clinicians about whether ethically relevant differences exist between withholding and withdrawing decisions.[63]

In the late twentieth century, about half of ICU deaths occurred after withholding or withdrawing LSTs.[64] Such decisions have increased over

---

[57] Stephen Scher and Kasia Kozlowska, *Rethinking health care ethics.*

[58] Sameera Karnik and Amar Kanekar, "Ethical issues surrounding end-of-life care: a narrative review."

[59] Gail Van Norman, "Ethical issues: withdrawing, withholding, and futility," 200-209.

[60] Ibidem.

[61] Ibidem.

[62] Randall Curtis and Jean-Louis Vincent, "Ethics and end-of-life care for adults in the intensive care unit," 1347-1353.

[63] Randall Curtis and Jean-Louis Vincent, "Ethics and end-of-life care for adults in the intensive care unit," 1347-1353. Gail Van Norman, "Ethical issues: withdrawing, withholding, and futility," 200-209.

[64] Thomas Prendergast, Michael Claessens, and John Luce, "A national survey of end-of-life care for critically ill patients," *American Journal of Respiratory and Critical Care Medicine* 158 (1998): 1163-1167.

time, and now precede death for 70 to 93 percent of patients dying in ICUs worldwide.[65] Studies indicate that a growing number of elderly patients across multiple cultures prefer not to have prolonged LSTs.[66] Older age is a significant factor for withdraw and withhold decisions by physicians and patients.[67]

In most Western countries, ethicists favour the view that doing and allowing are not sufficient, and may even be immaterial, in identifying morally relevant differences of withholding and withdrawing treatments.[68] The morally relevant features of withdrawing or withholding are the motives and intentions of reducing patient suffering and/or other burdens at the end of life and of respect for patient autonomy. Since these motives are the same with both decisions, there is widespread agreement that, in most cases, withdrawing and withholding are morally equivalent.[69] This position has been supported by numerous professional and international organizations (Table 1).[70]

Table 1. Some medical societies confirming ethical equivalency of withdrawing/withholding treatment decisions

American Medical Association
American College of Physicians
American College of Chest Physicians
American College of Critical Care Medicine
American Thoracic Society
The Australian and New Zeland Intensive Care Society
College of Intensive Care Medicine of Australia and New Zeland
Austrian Association of Intensive Care Medicine
Belgian Society of Intensive Care Medicine
Canadian Critical Care Society
The Duch Intensive Care Society
French Society of Intensive Care
The UK General Medical Council

[65] Gail Van Norman, "Ethical issues: withdrawing, withholding, and futility," 200-209.

[66] Elie Azoulay, Barbara Metnitz, Charles Sprung, Jean-François Timsit, François Lemaire, Peter Bauer, Benoît Schlemmer, Rui Moreno, Philipp Metnitz, and SAPS 3 investigators, "End-of-life practices in 282 intensive care units: data from the SAPS 3 database," *Intensive Care Medicine* 35 (2009): 623-630.

[67] Gail Van Norman, "Ethical issues: withdrawing, withholding, and futility," 200-9.

[68] Ibidem.

[69] Ibidem.

[70] Ibidem.

The ethical principle of "double effect" argues that the potential to hasten imminent death is acceptable if it comes as the known but unintended consequence of a primary intention to provide comfort and relieve suffering.[71] For example, it is acceptable to provide high doses of opioids if needed to control pain, even if there is the known and unintended effect of depressing respiration.[72]

Studies confirm that most disagreements between patients, families and clinicians can be resolved with good communication.[73] Additionally, open communication and respectful discourse between the physician and patient or their surrogates are prerequisites for resolving differences in values and finding an appropriate therapeutic pathway.[74]

Working with the dying requires tolerance of uncertainty, ambiguity, and existential challenges.[75] Clinicians must recognize and respect their own limitations and attend to their own needs in order to avoid being overburdened, overly distressed, or emotionally depleted.[76] Generally

---

[71] Michael Rabow, et at., "Palliative care and pain management," 72-99.

[72] Amy Kelley and Sean Morrison, "Palliative care for seriously ill," 747-755.

[73] Michael Rabow, et at., "Palliative care and pain management," 72-99.

[74] Gail Van Norman, "Ethical issues: withdrawing, withholding, and futility," 200-209.

[75] Sameera Karnik and Amar Kanekar, "Ethical issues surrounding end-of-life care: a narrative review," 24. Michael Rabow, et al., "Palliative care and pain management," 72-99.

[76] Michael Kearney, Radhule Weininger, Mary Vachon, Richard Harrison, and Balfour Mount, "Self-care of physicians caring for patients at the end of life," *The Journal of the American Medical Association* 301 (2009): 1155-1164. Madhav Goyal, Sonal Singh, Erica Sibinga, Neda Gould, Anastasia Rowland-Seymour, Ritu Sharma, Zackary Berger, Dana Sleicher, David Maron, Hasan Shihab, Padmini Ranasinghe, Shauna Linn, Shonali Saha, Eric Bass, and Jennifer Haythornthwaite, "Meditation programs for psychological stress and well-being," *JAMA Internal Medicine* 174 (2014): 357-368. Jennifer Wolff, Brenda Spillman, Vicki Freedman, and Judith Kasper, "A national profile of family and unpaid caregivers who assist older adults with health care activities," *JAMA Internal Medicine* 176 (2016): 372-379. Shannon Carson, Christopher Cox, Sylvan Wallenstein, Laura Hanson, Marion Danis, James Tulsky, Emily Chai, and Judith Nelson, "Effect of palliative care–led meetings for families of patients with chronic critical illness: a randomized clinical trial," *The Journal of the American Medical Association* 316 (2016): 51-62. Arif Kamal, Janet Bull, Steven Wolf, Keith Sweitz, Tait Shanafelt, Katherine Ast, Dio Kavalieratos, Christian Sinclair, and Amy Abernethy, "Prevalence and predictors of burnout among hospice and palliative care clinicians in the U.S," *Journal of Pain and Symptom Management* 51 (2016): 690-696. Dio Kavalieratos, Jennifer Corbelli, Di Zhang, Nicholas Dionne-Odom, Natalie Ernecoff, Janel Hanmer, Zachariah Hoydich, Dara Ikejiani, Michele Klein-Fedyshin, Camilla Zimmermann, Sally Morton, Robert Arnold, Lucas Heller, and Yael Schenker, "Association between palliative care and patient and caregiver outcomes.

speaking, clinicians must become proficient at delivering serious news and then dealing with its consequences (Table 2).[77]

Table 2. Suggestions for the delivery of serious news

Prepare an appropriate place and time.
Address basic information needs.
Be direct; avoid jargon and euphemisms.
Allow for silence and emotional ventilation.
Assess and validate patient reactions.
Respond to immediate discomforts and risks.
Listen actively and express empathy.
Achieve a common perception of the problem.
Reassure about pain relief.
Ensure follow-up and make specific plans for the future.

In this context, geriatrician David Dosa said: "In medical school, they used to teach you to be detached but empathetic when giving bad news. Listen and support but don't get involved. Easier said than done … But her eyes said something else. She knew there was a problem. People usually do… A doctor's office should feel like a safe place, an arena in which you can bring out your demons or your angels, your deepest fears and most intimate secrets…"[78]

The promise of non-abandonment is the central principle of end-of-life care and is a clinician's pledge to serve a caring partner, a resource for creative problem solving and relief of suffering, a guide during uncertain times, and a witness to the patient's experiences, no matter what happens.[79]

SPIRITUAL CONSIDERATIONS

Spirituality is the attempt to understand or accept the underlying meaning of life, one's relationships to oneself and other people, one's place in the universe, one's legacy, and the possibility of a "higher power" in the

A systematic review and meta-analysis," *The Journal of the American Medical Association* 316 (2016): 2104-2114.

[77] Michael Rabow, et al., "Palliative care and pain management," 72-99.

[78] David Dosa, *Making Rounds with Oscar: The Extraordinary Gift of an Ordinary Cat.*

[79] Michael Rabow, et al., "Palliative care and pain management," 72-99.

universe.[80] It may be distinguished from particular religious practices or beliefs.[81]

The end of life offers an opportunity for psychological, interpersonal and spiritual development.[82] Clinicians can inquire about a patient's spiritual concerns, and ask whether the patient wishes to discuss them.[83] Questions that might constitute an existential "review of systems" are presented in Table 3.[84]

Table 3. An existential review of systems

Intrapersonal
1. What does your illness/dying mean to you? 2. What do you think caused your illness? 3. How have you been healed in the past? 4. What do you think is needed for you to be healed now? 5. What is right with you now? 6. What do you hope for? 7. Are you at peace?
Interpersonal
1. Who is important to you? 2. To whom does your illness/dying matter? 3. Do you have any unfinished business with significant others?
Transpersonal
1. What is your source of strength, help, or hope? 2. Do you have spiritual concerns or a spiritual practice? 3. How does your spirituality relate to your illness/dying, and

[80] Tracy Balboni, George Fitchett, George Handzo, Kimberly Johnson, Harold Koenig, Kenneth Pargament, Christina Puchalski, Shane Sinclair, Elizabeth Taylor, and Karen Steinhauser, "State of the science of spirituality and palliative care. Research part II: screening, assessment and interventions," *Journal of Pain and Symptom Management* 54 (2017): 441-453. David Drummond and Lindsay Carey, "Assessing spiritual well-being in residential aged care: An exploratory review," *Journal of Religion and Health* 58 (2019): 372-390. Lindsay Carey and Timothy Hodgson, "Chaplaincy, spiritual care and moral injury: Considerations regarding screening and treatment," *Frontiers in Psychiatry* 9 (2018): doi:10.3389/fpsut.2018.00619. Kalli Stilos, Bill Ford, Tammy Lilien, and Jennifer Moore, "The role of spiritual care with the introduction of an end of life order set," *The Journal of Pastoral Care & Counseling* 73 (2019): 41-48.

[81] David Drummond and Lindsay Carey, "Assessing spiritual well-being in residential aged care: An exploratory review," 372-390.

[82] Ibidem.

[83] Ibidem.

[84] Michael Rabow, et al., "Palliative care and pain management," 72-99.

how can I help integrate your spirituality into your health care?
4. What do you think happens after we die?
5. What purpose might your illness/dying serve?
6. What do you think is trying to happen here?

Individuals may grow, even achieve a heightened sense of well-being or transcendence, in the process of dying.[85] Through listening, support and presence, clinicians may help foster this learning and be a catalyst for this transformation.[86]

Predictions concerning the end of life are particularly difficult and may lead to excessive or insufficient medical interventions.[87] Thus, geriatrician David Dosa described one case of death prediction from his practice: "There was the elderly woman who announced to me on 31 December 1999, that she had accomplished her objective of living to turn of the century. – *I am going to die today, Doctor*, – she told me quite casually. Every test showed she had nothing wrong… She simply came to the hospital because she was ready to die. As she foretold, she did die several hours later of unknown cases. Maybe some people just know when their time has come. Some cats, too."[88]

Oscar is a therapy cat living in the Steere House Nursing and Rehabilitation Center in Providence, Rhode Island, in the United States since 2005 (Fig. 1).[89]

He earned his fame through a publication in the New England Journal of Medicine in 2007.[90] Oscar was noted to make his own ward rounds in the nursing home, sniffing and observing patients.[91] Indifferent to most, he would then decide to curl up on the bed of only certain patients. The patients he identified would invariably die within a few hours. The cat was so precise in his assessment of impending death

[85] David Drummond and Lindsay Carey, "Assessing spiritual well-being in residential aged care: An exploratory review," 372-390.
[86] Michael Rabow, et al., "Palliative care and pain management," 72-99.
[87] Piotr Szawarski, "Classic cases revisited: Oscar the cat and predicting death," *Journal of the Intensive Care Society* 17 (2016): 341-345.
[88] David Dosa, *Making Rounds with Oscar: The Extraordinary Gift of an Ordinary Cat*.
[89] Wikipedia, *Oscar (therapy cat)*, available at http://en.wikipedia.org/wiki/Oscar _(therapy_cat) (accessed February 13, 2021).
[90] David Dosa, "A day in the life of Oscar the cat," *The New England Journal of Medicine* 357 (2007): 328-329.
[91] Piotr Szawarski, "Classic cases revisited: Oscar the cat and predicting death," 341-345. David Dosa, "A day in the life of Oscar the cat," 328-329.

that the staff developed a protocol that required that patient's family to be called in, in anticipation of death.[92] We read: "His mere presence at the bedside is viewed by physicians and nursing home staff as an absolute indicator of impending death."[93]

Fig. 1. Oscar (therapy cat)[94]

David Dosa later published a touching book about dementia – "Making Rounds with Oscar: The Extraordinary Gift of an Ordinary Cat."[95] Dr. Dosa reported about his first encounter with this famous therapy cat: "Making his way back up the hallway, Oscar arrives at Room 313. The door is open, and he proceeds inside. Mrs. K. is resting peacefully in her bed, her breathing steady but shallow. She is surrounded by photographs of her grandchildren and one from her wedding day. Despite these keepsakes, she is alone. Oscar jumps onto her bed and again sniffs the air. He pauses to consider the situation and then turns around twice before curling up beside Mrs. K. One hour passes. Oscar waits. A nurse walks into the room to check on her patient. She pauses to note Oscar's presence. Concerned, she hurriedly leaves the room and returns to her desk. She grabs Mrs. K.'s chart off the medical-records rack and begins to make phone calls. Within a half hour the family starts to arrive. Chairs are brought into the room, where the relatives begin their vigil. The priest is called to deliver last rites. And still, Oscar has not

---

92 Ibidem.

93 David Dosa, "A day in the life of Oscar the cat," 328-329.

94 Wikipedia, *Oscar (therapy cat)*, available at http://en.wikipedia.org/wiki/Oscar_(thera py_cat) (accessed February 13, 2021).

95 Piotr Szawarski, "Classic cases revisited," 341-345.

budged, instead it is purring and gently nuzzling Mrs. K. A young grandson asks his mother, ~What is the cat doing here?~ The mother, fighting back tears, tells him, ~He is here to help Grandma get to heaven.~ Thirty minutes later, Mrs. K. takes her last earthly breath. With this, Oscar sits up, looks around, then departs the room so quietly that the grieving family barely notices."[96]

While the reliability on the cat may seem strange in modern health care settings, the family of a patient that died with Oscar at the bedside reported: "It is not that we trusted the cat more than the nurse. Not, exactly. It was... well, there was just something about Oscar. He seemed so convinced of what he was doing. He was so clear in his intention and his dedication. This underscored the importance of the prediction of death and the acknowledgement of the dying process."[97] As of 2015, it is believed that Oscar has accurately predicted 100 deaths.[98] There is no solid scientific explanation for Oscar's behaviour. Hypotheses for this ability include that Oscar is picking up on the lack of movement in such patients, or that he can smell biochemicals released by dying cells.[99]

The simple example of a cat lying lovingly at a dying person's side calls health care practitioners back to the sacred role of being a compassionate presence.[100]

CONCLUSION

Much progress has been made to address the ethical issues and spiritual events surrounding end-of-life care. However, there is a necessity for further research into both topics, keeping in mind that "the experience of decision making can be a tremendous emotional burden likened to an arduous, unwelcome journey over unfamiliar territory filled with unrecognisable landmarks."

---

[96] David Dosa, "A day in the life of Oscar the cat," 328-329.

[97] David Dosa, *Making Rounds with Oscar: The Extraordinary Gift of an Ordinary Cat.*

[98] Wikipedia, *Oscar (therapy cat)*, available at http://en.wikipedia.org/wiki/Oscar_(therapy_cat) (accessed February 13, 2021).

[99] Ibidem.

[100] James Marroquin, "Book review-Making Rounds with Oscar: The Extraordinary Gift of an Ordinary Cat by David Dosa," *Proceedings (Baylor University Medical Center)* 26 (2013): 200-201.

ACKNOWLEDGEMENTS

The Ministry of Education and Science of the Republic of Serbia supported this article through grant N° 175015.

REFERENCES

Akhtar, Shamsuddin, and Rosenbaum Stanley. "Preface." In *Principles of Geriatric Critical Care*, eds. Shamsuddin Akhtar, and Stanley Rosenbaum. Cambridge: Cambridge University Press, 2018.

American Statistical Association. *World population likely to surpass 11 billion in 2100: US population projected to grow by 40 percent over next 85 years, 2015.* Available at www.sciencedaily.com/releases/2015/08/150810110634.htm (accessed February 13, 2021).

Azoulay, E., and Metnitz B., Sprung C., Timsit J.-F., Lemaire F., Bauer P., Schlemmer B., Moreno R., Metnitz P., SAPS 3 investigators. "End-of-life practices in 282 intensive care units: data from the SAPS 3 database." *Intensive Care Medicine* 35 (2009): 623-630.

Balboni, T., and Fitchett G., Handzo G., Johnson K., Koenig H., Pargament K., Puchalski C., Sinclair S., Taylor E., Steinhauser K. "State of the science of spirituality and palliative care. Research part II: screening, assessment and interventions." *Journal of Pain and Symptom Management* 54 (2017): 441-453.

Carey, L., and Hodgson T. "Chaplaincy, spiritual care and moral injury: Considerations regarding screening and treatment." *Frontiers in Psychiatry* 9 (2018): doi: 10.3389/fpsut.2018.00619.

Carey, L., and Hodgson T., Krikheli L., Soh R., Armour A.-R., Singh T., Impiombato C. "Moral injury, spiritual care and the role of chaplains: an exploratory scoping review of literature." *Journal of Religion and Health* 55 (2016): 1218-1245.

Carson, S., and Cox C., Wallenstein S., Hanson L., Danis M., Tulsky J., Chai E., Nelson J. "Effect of palliative care–led meetings for families of patients with chronic critical illness: a randomized clinical trial." *The Journal of the American Medical Association* 316 (2016): 51-62.

Curtis, R., and Vincent J.-L. "Ethics and end-of-life care for adults in the intensive care unit." *Lancet* 376 (2010): 1347-1353.

Dosa, D. "A day in the life of Oscar the cat." *The New England Journal of Medicine* 357 (2007): 328-329.

Dosa, D. *Making Rounds with Oscar: The Extraordinary Gift of an Ordinary Cat.* New York: Hyperion, 2010.

Drummond, D., and Carey L. "Assessing spiritual well-being in residential aged care: An exploratory review." *Journal of Religion and Health* 58 (2019): 372-390.

Fink-Samnick, E. "The evolution of end-of-life care. Ethical implications for case management." *Professional Case Management* 21 (2016): 180-192.

Glascock, A., and Feinman, S. "A holocultural analysis of old age." *Comparative Social Research* 3 (1980): 311-332.

Goyal, M., and Singh S., Sibinga E., Gould N., Rowland-Seymour A., Sharma R., Berger Z., Sleicher D., Maron D., Shihab H., Ranasinghe P., Linn S., Saha S., Bass E., Haythornthwaite J. "Meditation programs for psychological stress and well-being." *JAMA Internal Medicine* 174 (2014): 357-368.

Gusmano, M., and Allin S. "Framing the issue of ageing and health care spending in Canada, the United Kingdom and the United States." *Health Economics, Policy and Law* 9 (2014): 313-328.

Huynh, T., and Kleerup E., Wiley J., Savitsky T., Guse D., Garber B., Wenger N. "The frequency and cost of treatment perceived to be futile in critical care." *JAMA Internal Medicine* 173 (2013): 1887-1894.

Kamal, A., Bull, J., Wolf, S., Sweitz, K., Shanafelt, T., Ast, K., Kavalieratos, D., and Sinclair C., Abernethy A. "Prevalence and predictors of burnout among hospice and palliative care clinicians in the U.S." *Journal of Pain and Symptom Management* 51 (2016): 690-696.

Karnik, S., and Kanekar, A. "Ethical issues surrounding end-of-life care: a narrative review." *Healthcare (Basel)* 4 (2016): doi: 10.3390/healthcare4020024.

Kavalieratos, D., and Corbelli J., Zhang D., Dionne-Odom N., Ernecoff N., Hanmer J., Hoydich Z., Ikejiani D., Klein-Fedyshin M., Zimmermann C., Morton S., Arnold R., Heller L., Schenker Y. "Association between palliative care and patient and caregiver outcomes. A systematic review and meta-analysis." *The Journal of the American Medical Association* 316 (2016): 2104-2114.

Kearney, M., and Weininger R., Vachon M., Harrison R., Mount B. "Self-care of physicians caring for patients at the end of life." *The Journal of the American Medical Association* 301(2009): 1155-1164.

Kelley, A., and Morrison S. "Palliative care for seriously ill." *The New England Journal of Medicine* 373 (2015): 747-755.

Marroquin, J. "Book review-Making Rounds with Oscar: The Extraordinary Gift of an Ordinary Cat by David Dosa." *Proceedings (Baylor University Medical Center)* 26 (2013): 200-201.

Mohanti, B. "Ethics in palliative care." *Indian Journal of Palliative Care* 15 (2009): 89-92.

Nelson-Becker, H., and Ai A., Hop, F., McCormic, T., Schluete, J., Cam, J. "Spirituality and religion in end-of-life care ethics: the challenge of interfaith and cross-generational matters." *The British Journal of Social Work* 45 (2015): 104-119.

Oliveira, Kristin and Maerz Linda. "Epidemiology of critical illness in the elderly." In *Principles of Geriatric Critical Care,* eds. Shamsuddin Akhtar, and Stanley Rosenbaum, 1-10. Cambridge: Cambridge University Press, 2018.

Prendergast, T., and Claessens M., Luce J. "A national survey of end-of-life care for critically ill patients." *American Journal of Respiratory and Critical Care Medicine* 158 (1998): 1163-1167.

Rabow, Michael, and Pantilat Steven, Steiger Scott, Naidu Ramana. "Palliative care and pain management." In *Current Medical Diagnosis and Treatment*, eds. Maxine Papadakis, and Stephen McPhee, 72-99. New York: Mc Graw Hill, 2018.

Scher, Stephen, and Kozlowska Kasia. *Rethinking health care ethics eBook*. doi: 10.1007/981-13-0830-7. New York: Palgrave Pivot, 2018.

Shah, E. "The ageing and development report 1999: poverty, independence and the world`s older people." *British Medical Journal* 321 (2000): 517.

Stilos, K., and Ford B., Lilien T., Moore J. "The role of spiritual care with the introduction of an end of life order set." *The Journal of Pastoral Care & Counseling* 73 (2019): 41-48.

Szawarski, P. "Classic cases revisited: Oscar the cat and predicting death." *Journal of the Intensive Care Society* 17 (2016): 341-345.

United Nations Department of Economic and Social Affairs. *World population prospects: the 2015 revision*, 2015. Available at http://esa.un.org/unpd/wpp/ (accessed February 13, 2021).

Van Norman, Gail. "Ethical issues: withdrawing, withholding, and futility." In *Principles of Geriatric Critical Care*, eds. Shamsuddin Akhtar, and Stanley Rosenbaum, 200-209. Cambridge: Cambridge University Press, 2018.

Wikipedia. *Oscar (therapy cat)*. Available at http://en.wikipedia.org/wiki/Oscar_(therapy_cat) (accessed February 13, 2021)

Wolff, J., and Spillman B., Freedman V., Kasper J. "A national profile of family and unpaid caregivers who assist older adults with health care activities." *JAMA Internal Medicine* 176 (2016): 372-379.

World Health Organization. *Global health observatory data: life expectancy*, 2013. Available at www.who.int/gho/mortality_burden_disease/life_tables/situation_trends_text/en/ (accessed February 13, 2021).

THE LACK OF DENTAL ETHICS EDUCATION IN THE UNDERGRADUATE CURRICULA IN DENTAL SCHOOLS OF SERBIAN UNIVERSITIES

Jelena Roganović[1]

INTRODUCTION

Although dental ethics has already established itself as an important and independent discipline structured on the contemporary problems in dentistry, one can still hear the opinion that teaching ethics is not necessary since it is nothing more than a matter of common sense and conscience. In clinical practice, however, common sense is not a universal "phenomenon," and there are plenty of occasions where conscience is not enough, and doctors could put themselves into serious legal difficulties[2]

Why is studying dental ethics important nowadays?

Nowadays, more than ever, there is a necessity for guidelines in dental ethics in order to point at the ethical responsibility of everyone working in the dental practice towards patients, each other, as well as towards actions in dental research. In the contemporary dental clinical practice, aesthetic dentistry became a significant branch leading to several ethical issues arising since aesthetic dentistry aims to improve the patient's

---

[1] Department of Pharmacology in Dentistry, School of Dental Medicine, University of Belgrade, Belgrade, Serbia.

[2] Alison Patrick, "A review of teaching ethics in the dental curriculum: challenges and future developments," *European Journal of Dental Education* 21(2017): e114-e118.

physical appearance without targeting functionality. Aesthetic procedures may involve significant and irreversible harm to the orofacial tissues. For instance, up to 30% of sound hard tissue may be removed for a porcelain veneer preparation, while around 70% of sound tooth structure could be removed during preparation for ceramic crowns in anterior teeth[3]. Furthermore, a doctor of dental medicine has to clearly differentiate between the health care needs and desires of the patient, make this difference clear to the patient, and work in the best interest of the patient. One should admit however, that in reality, dentistry is a business. Therefore, there is a conflict of interest because the dentist is a doctor, providing health care but also a professional in charge of private practice, tempted to gain more money by recommending a particular expensive procedure, regardless of the best interest of the patient. Noteworthy, not so long ago, it was believed that dentists' integrity and professionalism could not be associated with advertising campaigns. Nowadays, advertising is no longer considered unethical but rather useful in the promotion of oral health and preventive programs. In the light of contemporary ethical demands and challenges, American as well as European Dental Education Associations recognized the necessity of preparing dental students for solving the ethical dilemmas in order to develop competence and to apply ethical and legal standards in the daily dental practice.

Is there a difference between medical and dental ethics?

The great differences between medicine and dentistry lay in the professional goals as well as in the patient's expectations. While professional goals in medicine aim at health per-se, dentistry has to aim also at the patient's psychological being which include the patient's desires. In line with this, the medical doctor is expected to "care" about health while the dentist needs to "satisfy" the patient. Therefore, in medicine, patients are grateful if medical success is achieved, while in dentistry, a patient's dissatisfaction could not be related to oral health issues but rather to a disappointment of a "client" regarding his/her desires fulfilment. Furthermore, the dental patient is often obligated to "buy" dental works such as dental fillings, prosthetic dentures or dental

---

[3] Daniel Edelhoff, John Sorenson, *"Tooth structure removal associated with various preparation designs for anterior teeth," Journal of Prosthetic Dentistry* 87 (2002): 502–509.

crowns, which additionally complicates the situation regarding the patient's dissatisfaction in the dental office.[4]

Regarding legal and practical responsibility, while the responsibility of the medical doctor finishes when he refers the patient to another specialist or health worker, in the dental office, especially in the private clinic, the responsibility is entirely on the dentist. The dental technician, dental hygienist, dental assistants and even anaesthesiologist work under the supervision and direction of the dentist, and the dentist is responsible for their work, even if they are not employed in that particular dental office. In the light of these differences, it seems that ethical demands in medicine and dentistry are not similar.

Nevertheless, the legal liability of medical and dental doctors to their patients is based on the same principles and many ethical issues are identical. For instance, any dental or medical treatment requires informed consent from the patient or legal guardian. Professional confidentiality and respect of the patient's privacy, loyalty to the patient and equality as principal behaviour, comprise some of the most important ethical principles in both medicine and dentistry. Also, medical and dental practices are obliged to keep medical or dental records as well as to obey to legal regulative regarding medical and dental certificates. The patients' rights do not discriminate between medicine and dentistry. Laws are dealing with legal issues arising in all kind of health care institutions and various health care professionals.[5]

DENTAL ETHICS IN CLINICAL PRACTICE

Due to a complex dentist-patient relationship, there are constant and daily, demanding ethical challenges in front of a doctor of dental medicine. In the light of the basic principles of bioethics, including autonomy, beneficence, non-maleficence and justice, the responsibility of the dentist remains demanding.

The principle of autonomy, the patient's right to make a decision independently regarding health, could be challenged in the following situations:

---

[4] David Ozar, "Professionalism: challenges for dentistry in the future," *The Journal of Forensic Odonto-stomatology* 30 Suppl 1 (2012):72-84.

[5] Ibidem.

- When the patient is not sufficiently informed and has no complete understanding of the intervention due to inadequate explanation by the doctor
- When the patient does not understand information due to personal incompetence
- When the personal state, such as intensive pain, influence the patient's understanding of the given information

These examples suggest that an autonomous patient can make a non-autonomous decision and that this principle should be considered an autonomous act rather than an autonomous patient. The dentist is obligated to fully inform the patient about the risks and benefits of each proposed treatment but leaving the patient to make the final decision on the treatment preferred. The patient has a right to truthful information on his/her diagnosis, prognosis and therapy, with all benefits and risks associated with the suggested procedure. However, in contemporary dental medicine, informing the patient may be questionable since a wide range of materials and techniques could be used. It is ethically disputable to recommend a specific expensive procedure, despite the best interest of the patient, in order to gain some profit. In line with this, the involvement of the patient's capacity of payment should be considered; however, it could not be a relevant criterion for proposing "best treatment," especially in the private sector, where most dentists practice.[6] One should perform procedures with minimal risk and maximum benefit by choosing the minimally invasive procedures, balancing between aesthetics and functioning.

Even if a patient insists, the dentist should not perform certain procedures if that could harm the patient and compromise his/her professional reputation. On the other side, sometimes, the doctor has to respect the patient's decision even if it is not the best solution.

Another major issue includes the refusal to provide dental care to certain patients on discriminatory grounds (e.g. HIV positive status). The justice or justification as an ethical principle holds on the respect of all patients equally, regardless of their gender, nationality, religion or health status. The doctor should always act in the manners of honesty and empathy and give patients what they deserve and have the right on. Principles of confidentiality and loyalty comprise preservation of all

[6] Ibidem.

information regarding the patient's state and treatment along with doing best for the patient with respect to him/her as a person.

DENTAL ETHICS IN THE RESEARCH

Medical and dental research has evolved during the past years in Serbia into fast-developing academic activities aimed to improve both diagnostic and therapeutic procedures. Since medical and dental research involve human participants, these actions must be regulated in order to protect the participants' rights. Namely, clinical trials that involve human participants need to be guided by fundamental ethical principles to ensure the protection of the safety, welfare and autonomy of research participants.[7] Nevertheless, animal studies also need to be guided by ethical principles based on the protection of animals' welfare. In line with these, international standards mandate the necessary review of research designs and projects by Research Ethics Committees (RECs). In Serbia, institutions where research is done, have their own Ethical committees licensed to review and approve research activities, whether involving human participants or animals.[8] Dental research is committed to the investigation of the causes and treatment of dental and oral diseases and adheres to the same ethical standards as those valid in medicine. The importance of studying dental research ethics has been emphasized by the studies showing several scientific misconducts in areas of dental research.[9] Supporting the necessity of dental research ethics continuing learning, surveys among academics show large discrepancies between theoretical and practical knowledge. Namely, in a study by El-Dessouyki and colleagues it was shown that almost 30% of the respondents thought that research performed on blood samples does not need informed consent from the patient. In the same study, almost half of the academic researchers believed that vulnerable subjects (e.g., children and the mentally ill) could provide informed consent to participate in research.[10]

[7] Zorica Vučinić, Ljiljana Djukić, "The Regulation of Clinical Trials in the Republic of Serbia," *Hospital Pharmacology* 2 (2015): 322-327.

[8] Ibidem.

[9] Muriel Bebeau, Elizabeth Davis, "Survey of ethical issues in dental research," *Journal of Dental Research* 75 (1996): 845-855

[10] Hadir El-Dessouky, Amr Abdel-Aziz, Chadi Ibrahim, Moni Malini, Reham Abul Fadl, Henry Silverman, "Knowledge, Awareness, and Attitudes about Research Ethics among

Accordingly, the authors suggested compulsory training in ethics for academics, especially regarding informed consent issues, including those issues associated with the sensitive population of the research participants.

Regarding another aspect of research, Eastwood and colleagues found that a small percentage of survey participants- postdoctoral research fellows (2%), was willing to fabricate data for a grant application or a paper, while a significant percentage of participants (27%) was willing to select or omit data to improve the results.[11]

A meta-analysis of surveys involving scientists' self-report of research misconduct showed that 0.3% to 4.9% of the scientists admitted that they had fabricated or falsified data at least once.[12] However, one could see that these numbers of those who "did it" are nevertheless lower than those who "would be willing to do," suggesting the difference between "perception" and actual "practice." Finally, researchers recommend the development of education in research ethics at universities, emphasizing the significance of vulnerable participants' consideration, responsible research conduction as well as clarification of Research Ethical Committees involvement.

DENTAL ETHICS AND SERBIAN UNIVERSITIES

The growing understanding of the nature and requirements of professionalism in dental medicine is changing the conception of dentistry, considering it as both health care and research activities. While the association between ethics and patient care is an obvious and important one, ethical behaviour is also necessary even before clinical experiences. Namely, there is a necessity to "incorporate" ethical behaviour among dental students in ways to address inappropriate ethical conduct among students and graduates. This way, teaching medical and dental ethics evolves with contemporary and acute topics such as cheating, alcohol use or sexual attraction to patients. In large, nowadays,

---

Dental Faculty in the Middle East: A Pilot Study," *International Journal of Dentistry* 2011 (2011): 694759.

[11] Susan Eastwood, Pamela Derish, Evangeline Leash, Stephen Ordway, "Ethical issues in biomedical research: perceptions and practices of postdoctoral research fellows responding to a survey," *Science and Engineering Ethics* 2 (1996): 89-114.

[12] Daniele Fanelli, "How many scientists fabricate and falsify research? A systematic review and meta-analysis of survey data," *PLoS One* 4 (2009): e5738.

it seems to be important to include dental ethics education in the undergraduate curricula in order for every dental graduate to adopt ethics as a responsibility towards providing the highest standards of dental care. In line with this, the Association for Dental Education in Europe (ADEE), as well as the American Dental Association (ADA), defined the necessary skills that a dentist should possess, including those that include ethical and professional behaviour. Noteworthy, Patrick[13] pointed that the increased number of dental schools that included a course of ethics in dental curricula most certainly reflects the understanding that such practice resulted in producing well-rounded and professional dentists. In North America, from 91% of responding dental schools, which included at least one ethics course within their curriculum, 61 % of schools reported teaching ethics in the first year while around 40% in third and fourth years. Courses mainly were found to range from one to three credits. Odom and colleagues emphasize the need for an ethics course to begin in the first year in order to be applicable during student training in the next years.[14] Noteworthy, although dramatic differences exist between schools regarding teaching hours and course content, positive results were evident.

The dental ethics education in Universities of Serbia is marginalized. None of the current curricula at Serbian Universities recognize dental ethics as a separate course. In the current curricula, the University of Novi Sad has medical ethics (and not dental, although there is evidence supporting that dental ethics faces different ethical issues) as a compulsory subject on the first year of undergraduate dentistry studies with a total of 45 hours. Dentistry studies at the Universities of Niš do not offer a separate course in ethics; instead, ethical issues are considered under the course "Medicine and society" with a total of 30 hours in the first semester of the studies. The University of Kragujevac recognizes ethics (but not dental ethics) as a compulsory subject in the first year of studies with a total of 60 hours. Surprisingly, the Dental school at the University of Belgrade does not recognize ethics as a course. Considering educational methods, lectures *ex- cathedra* is predominantly present (Table

[13] Alison Patrick, "A review of teaching ethics in the dental curriculum," e114-e118.

[14] John Odom, Phyllis Beemsterboer, Ted Pate, Karl Haden, "Revisiting the status of dental ethics instruction," *Journal of Dental Education* 64 (2000): 772-774.

1). One could easily conclude that this marginalization of dental ethics education could result in students perceiving ethics as unimportant.

Table 1. Courses of ethics in undergraduate curricula in Dental schools at Universities in Serbia

	University of Belgrade	**University of Novi Sad**	**University of Niš**	**University of Kragujevac**
Course in ethics	none	Medical ethics and sociology	Medicine and society	Ethics
Obligatory/ elective		obligatory	obligatory	obligatory
Year of studies		first	first	first
Number of hours		45	30	60
ESPB		3	2	4
Educational method		lectures	lectures	lectures

DENTAL ETHICS CURRICULUM

Today, contemporary teaching methods emphasize the importance of a more interactive mode that promotes introspection and self-knowledge, components of emancipatory and practical knowledge. Noteworthy, teaching ethics shows that advantage has to be given to practical and emancipatory knowledge instead of technical and scientific knowledge.[15] Namely, practical knowledge recognizes communication and group problem-solving behaviours, while emancipatory knowledge recognizes self-reflection and ethical inspection. Results point that successful methods of ethics teaching are based on the formation of these forms of knowledge.[16]

[15] Joan Whipp, Donald Ferguson, Linda Wells, Anthony Iacopino, "Rethinking knowledge and pedagogy in dental education," *Journal of Dental Education* 64 (2000): 860-866.

[16] Nancy Berk, "Teaching ethics in dental schools: trends, techniques, and targets," *Journal of Dental Education* 65 (2001): 744-750.

Regarding optimal dental ethics curriculum, the recommendation is to create a balance between basic ethical and dental professional knowledge and behaviour by combining different methodological approaches.

Case-based learning has been strongly promoted in medical school ethics instruction. Namely, Perkins et al. suggested that teaching medical ethics should be via realistic patient cases.[17] Indeed, many dental schools have endorsed this methodology showing its efficacy by means of capturing the students' attention and making ethics instruction more realistic and clinically applicable. Furthermore, the small-group learning format showed optimal results in the development of the student-instructor relationship and introspection. More importantly, this way, educators had an opportunity to serve as role models for their students regarding obeying ethical standards and behaviour.[18] Medical ethics courses utilized problem-based learning (PBL) groups, which could be a successful approach for learning ethics in dental education. This should be associated with a narrative approach (e.g., stories) that could generate the optimal setting for student reflection. The use of the narrative approach enables the consideration of the overall context of ethical dilemmas in the clinical settings as well as consideration of the complex doctor-dental staff, or doctor-patient relationships and emotions.[19] Particular attention should be put on the less obvious and unintentional messages sent by dental educators. Namely, these messages may have an important impact on the ethics experience as well, determining its significance. It is noteworthy that dental ethics educators should be aware of what has been said between the lines: if we close our eyes to exams cheating, if we do not penalize plagiarism or inappropriate behaviour towards colleagues or patients, what message do we send to our students?

On the other hand, in order to take an exam or finish the semester, students need to fulfil clinical requirements ultimately, and sometimes, that means that they will do "anything," including some ethical violations, to do the assignments. Do we then contribute to the ethical

[17] Henry Perkins, Cynthia Geppert, Helena Hazuda, "Challenges in teaching ethics in medical schools," *American Journal of Medical Sciences* 319 (2000): 273-8.

[18] Donald Dibbern, Eric Wold, "Workshop-based learning: a model for teaching ethics," *The Journal of the American Medical Association* 274 (1995): 770-771.

[19] Anne Jones, "Narrative in medical ethics," *British Medical Journal* 318 (1999): 253-256.

dilemmas faced by students, just allowing them to do so? These "hidden" messages could certainly diminish ethics teaching efforts. The curricula of clinical subjects in Dental school in Belgrade comprise numerical tasks in order to achieve clinical competency (and graduation). This way, it is associated with the risk of the patient being perceived as intervention or number and of students putting first interventions and deadlines instead of the patient's needs.

Thus, it seems that the role of the school has to be undoubtfully to obey and respect ethical grounds which proclaims. The major goal has to be optimal patient care, and therefore, mentors should be involved to track progress in clinical and ethical behaviour.

Altogether, evidence suggests that dental education should be a separate course, best to start before clinical training, while educational methods must include interactive mode with incorporated case studies as a method of teaching. The students should be engaged in lively classroom discussions regarding issues and cases that they comprehend as realistic and relevant to their current and future dental and scientific practice. There is also consensus among students regarding curriculum: intensive ethics course with small group format instead of exclusively didactic, involving review of everyday ethical dilemmas as well as ethical behaviour of mentors and superiors.[20] Regarding the clinical evaluation of students during course, ratings of student-patient interactions is the preferable method. Also, the educators as moderators could provide a forum for examining and discussing ethical issues with students. Nevertheless, we should not forget that as educators, we model the ethical behaviour of students, which adopt the standards of honesty and integrity while in dental school. Furthermore, teaching ethics should include common as well as not-so-common ethical issues encountered during everyday practice. Until recently, there was a suggestion that ethics teaching should be focused on the unusual and not-so-common cases of ethical dilemmas, but Truog and colleagues pointed to an often-missed consequence of mentioned approach- focusing on rare and controversial cases in order to demonstrate ethical principles. Namely, these authors suggested that by focusing on the not common cases, there is a risk of

[20] Thorsten Langer, Danny Jazmati, Ole Jung, Christian Schulz, Martin Schnell, "Medical Students' Development of Ethical Judgment - Exploring the Learners' Perspectives using a mixed methods approach," *GMS Journal for Medical Education* 33 (2016): Doc74.

neglecting a number of ethical decisions encountered during the daily clinical practice.[21]

Ethics learning should be continued after graduate studies also, since continuing education in professional as well as in ethical behaviour is necessary for maintaining competency, regardless of the student's knowledge and skills. For example, Robin and Caniano found that while most paediatric surgery residents in the survey reported that they felt competent to handle conflicts in clinical practice, almost half of them did not accurately answer questions regarding usual ethical dilemmas involving children.[22]

A course of medical ethics was not a part of the residency curriculum, although survey results support the efforts to continue with ethics education in the postgraduate period in order to refresh and widen the knowledge based on undergraduate studies. The tracking of the students' academic and practical activities might also provide valuable information on ethics educational programmes efficacy. Namely, it still is hard to validate ethics teaching outcomes, so assessments of ethical behaviour, including violations, could provide us with some insight into the impact of ethics learning on the ethical behaviour of students.

CONCLUSION

Although an excellent student in dental ethics does not necessarily mean an ethical dentist, a well-designed program of dental ethics education should provide clear principles and practical guidance on how to apply basic ethical principles of compassion, kindness and integrity into clinical and/or scientific practice. Dental ethics education adheres to fundamental academic values and aims to foster pride and honour in the dental profession.

---

[21] Robert Truog, Stephen Brown, David Browning, Edward Hundert, Elizabeth Rider, Sigall Bell, Elaine Meyer, "Microethics: the ethics of everyday clinical practice," *Hastings Center Report* 45 (2015): 11–17.

[22] Michael Robin, Donna Caniano, "Analysis of clinical bioethics teaching in pediatric surgery residency," *Journal of Pediatric Surgery* 33 (1998): 373-377.

REFERENCES

Bebeau, Muriel and Davis Elizabeth. "Survey of ethical issues in dental research." *Journal of Dental Research* 75 (1996): 845-855.

Berk, Nancy. "Teaching ethics in dental schools: trends, techniques, and targets." *Journal of Dental Education* 65 (2001): 744-750.

Dibbern, Donald, and Wold, Eric. "Workshop-based learning: a model for teaching ethics." *The Journal of the American Medical Association* 274 (1995): 770-771.

Eastwood, Susan, and Derish Pamela, Leash Evangeline, Ordway Stephen. "Ethical issues in biomedical research: perceptions and practices of postdoctoral research fellows responding to a survey." *Science and Engineering Ethics* 2 (1996): 89-114.

Edelhoff, Daniel, and Sorenson John. *"Tooth structure removal associated with various preparation designs for anterior teeth."* *Journal of Prosthetic Dentistry* 87 (2002): 502–509.

El-Dessouky, Hadir, and Abdel-Aziz Amr, Ibrahim Chadi, Moni Malini, Abul Fadl Reham, Silverman Henry. "Knowledge, Awareness, and Attitudes about Research Ethics among Dental Faculty in the Middle East: A Pilot Study." *International Journal of Dentistry* 2011 (2011): 694759.

Fanelli, Daniele. "How many scientists fabricate and falsify research? A systematic review and meta-analysis of survey data." *PLoS One* 4 (2009): e5738.

Jones, Anne H. "Narrative in medical ethics." *British Medical Journal* 318 (1999): 253-256.

Langer, Thorsten, and Jazmati Danny, Jung Ole, Schulz Christian, Schnell Martin. "Medical Students' Development of Ethical Judgment - Exploring the Learners' Perspectives using a mixed methods approach." *GMS Journal for Medical Education* 33 (2016): Doc74.

Odom, John, and Beemsterboer Phyllis, Pate Ted, Haden Karl. "Revisiting the status of dental ethics instruction." *Journal of Dental Education* 64 (2000): 772-774.

Ozar, David. "Professionalism: challenges for dentistry in the future." *The Journal of Forensic Odonto-stomatology* 30 Suppl 1 (2012): 72-84.

Patrick, Alison. "A review of teaching ethics in the dental curriculum: challenges and future developments." *European Journal of Dental Education* 21(2017): e114-e118.

Perkins, Henry, and Geppert Cynthia, Hazuda Helen. "Challenges in teaching ethics in medical schools." *American Journal of Medical Sciences* 319 (2000): 273-8.

Robin, Michael L., and Caniano Donna A. "Analysis of clinical bioethics teaching in pediatric surgery residency." *Journal of Pediatric Surgery* 33 (1998): 373-377.

Truog, Robert, and Brown Stephen, Browning David, Hundert Edward, Rider Elizabeth, Bell Sigall, Meyer Elaine. "Microethics: the ethics of everyday clinical practice." *Hastings Center Report* 45 (2015): 11–17.

Vučinić, Zorica, and Djukić LJiljana. "The Regulation of Clinical Trials in the Republic of Serbia." *Hospital Pharmacology* 2 (2015): 322-327.

Whipp, Joan, and Ferguson Donald, Wells Linda, Iacopino Anthony. "Rethinking knowledge and pedagogy in dental education." *Journal of Dental Education* 64 (2000): 860-866.

THE CHALLENGES OF THE DOUBLE ROLE OF THE ETHICS COMMITTEE AT THE FACULTY OF MEDICINE FROM THE UNIVERSITY OF BELGRADE

Jelena Šantrić[1]

INTRODUCTION

The Faculty of Medicine of the University of Belgrade is the oldest and largest faculty of medicine in Serbia, and among the largest medical educational institutions in the region.

In 2020, the Faculty will celebrate 100 years from the first lecture delivered by the professor of anatomy Niko Miljanic, which led to the first generation of medical students in Serbia.

The most important resources of the Faculty are its people: about 3000 undergraduate students, over 5530 postgraduate students, over 700+ academic teaching staff and 300+ clinical junior teaching staff, 270+ non-teaching staff, using more than 60.000 m2 of the space at the faculty´s buildings and clinics and hospitals as teaching bases.[2] The size of the Faculty reflects the ramified and all-encompassing activities in the field of higher education, research and health services carried by the Faculty and its employees.

The current local Law on Higher Education proposes the role of ethics committees at universities and faculties in protecting academic integrity by conducting the procedure in cases of violation of the Code

[1] Department of Faculty Services, Faculty of Medicine, University of Belgrade, Serbia.

[2] Data from Faculty Services, Faculty of Medicine, University of Belgrade, available at http://med.bg.ac.rs/?page_id=14465&lang=en (accessed January 20, 2021).

of Professional Ethics adopted by the National Council for Higher Education. These violations concern non-academic behaviour in teaching and research work at universities, as well as in publishing from teaching staff, researchers and students, with particular emphasis on plagiarism, false authorship, falsification of results and auto-plagiarism.

In a similar way, the Law on Health Care proposes the role of an ethical board in health care institutions in protecting the regulations of professional ethics in the health care practice clinical research, research in public health, as well as in taking and using human organs, cells and tissues from living and deceased donors.

The ethics committee also has an important role in improving the overall culture in biomedical research and communication between researchers, institutions and society, and consequently in improving the quality of the health care system. Therefore, the challenges of the dual role of the Ethics Committee at the Faculty of Medicine from the University of Belgrade will be discussed.

ETHICAL PRINCIPLES IN HIGHER EDUCATION

The Law on Higher Education,[3] one of the main principles in higher education, implies academic integrity and unity of teaching, research and professional work. Academic integrity is based on preserving the dignity of the profession, improving moral values, protecting the value of knowledge, raising awareness of the responsibility of all members of the academic community and promoting human rights and freedoms.

Academic freedoms, according to the Law,[4] include the freedom of scientific and artistic research, freedom to publish and publicly present scientific results and artistic achievements, alongside respecting intellectual property rights.

The professional body[5] of the independent higher education institution adopts a code of professional ethics and rules of conduct in the institution, which establishes the ethical principles for publishing scientific and artistic results, intellectual property, relations between

[3] Law on Higher Education (Official Gazette RS 88/17, 27/18 - an. law, 73/18), Article 7, Available at: https://www.nat.rs/wp-content/uploads/2019/01/Zakon-o-visokom-obrazovanju-2017-ENGL.pdf (accessed January 20, 2021).

[4] Ibidem, Article 5.

[5] Ibidem, Article 65.

teachers and associates, other employees and students, procedures in the performance of the higher education institution and teachers, legal transactions, as well as in relation to the public and the media.

The National Council for Higher Education, established to ensure the development and improvement of the quality of higher education, provides the basis for a code of academic integrity and conflict of interest in the election of a professor or associate.

The Code of Academic Integrity in Higher Education Institutions in the Republic of Serbia implies that the members of the academic community are obliged to preserve the integrity of the higher education system and of the higher education institution itself. They are also obliged to meet their responsibilities towards students, colleagues and employees of the higher education institution and within the academic community responsibly, conscientiously and professionally.

The members of the academic community guarantee the originality of the work they publish, as well as the accuracy in presenting and providing information on the origin of ideas and statements used in the work. Plagiarism, false authorship, fabrication and falsification of results and auto plagiarism are not permitted in scientific and professional work. These prohibitions apply to all written works of teaching and scientific staff and students (seminar paper, master's thesis, specialist thesis, doctoral dissertation, journal work, textbook chapter, chapters in proceedings etc.).[6]

The members of the academic community are also responsible for violating these prohibitions in case of unconscious and irresponsible presentation or neglecting the facts, in their teaching or scientific research work (in the papers for the evaluation of a doctoral dissertation, election to the title of a teacher or an associate).

A special responsibility lies in the hands of the supervisor in the preparation of the doctoral dissertation, as well as the instructor who guides the student in the preparation of other written papers.

The process of detecting non-academic behaviour is clearly regulated. Different bodies are involved in the procedure, such as the ethics committee of the faculty, the scientific council of the faculty, the expert

[6] Law on Copyright and Related Rights (Official Gazette RS 104/2009, 99/2011, 119/2012 and 29/2016) Available at: http://www.zis.gov.rs/upload/THE%20LAW%20ON%20COPYRIGHT%20AND%20RELATED%20RIGHTS-version%20in%20force%204.1.2013._.pdf (accessed January 20, 2021).

committee (which is formed by the scientific council) and the second-instance body for professional ethics at the university level.

As a violation of the Code, the following could be pronounced:

- measures for established non-academic behaviour in the form of plagiarism, false authorship, fabricating and falsifying results and self-promotion such as advertising diploma on completion of the scientific name PhDs null and void or revocation of teaching or associate titles;
- measures for minor injuries identified or negligent treatment of the obligation to correct academic behaviour such as warning, public warning and public condemnation.

ETHICAL PRINCIPLES IN HEALTHCARE

The Law on Healthcare[7] predicts that healthcare professionals perform healthcare activities in accordance with professional standards, approved good practice guides, treatment protocols and a code of professional ethics. Healthcare professionals assume professional, ethical, criminal, material and disciplinary responsibility for their work, in accordance with the Law.

Upon graduating, health professionals sign a declaration - an oath, that in carrying out their vocation they will adhere to the principles set out in the Hippocratic Oath, as well as the principles of professional ethics.

In the Republic of Serbia, the Government appoints and dismisses the President and members of the Ethics Committee of Serbia,[8] at the proposal of the Minister. The Ethics Committee of Serbia is a professional body that takes care of the provision and implementation of health care in accordance with the principles of professional ethics, the principles of respect for human values and the rights of the child.

The Ethics Committee of Serbia consists out of ten members who are selected among experts with significant results in their work, as well as contributions in the field of healthcare, professional ethics of health workers and humanities - five of whom are chairmen of the ethics

[7] Law on Healthcare (Official Gazette RS 25/19) Article 155, Available at: http://www.skriningsrbija.rs/files/File/English/Republic_of_Serbia_Healthcare_Law.pdf (accessed January 20, 2021).

[8] Ibidem, Articles 141-142.

committees of the competent chambers of health workers, and one with a law degree and clinical trial experience.

Among the numerous responsibilities of the Ethics Committee of Serbia, the following ones are listed as crucial:

- to predict the basic principles of professional ethics of health workers and health associates and monitor their application in performing health care activities on the territory of Serbia;
- to propose a code of conduct for employees in the health care system and monitor its implementation on the territory Serbia;
- to coordinate the work of ethics committees in healthcare institutions;
- to monitor the implementation of scientific, medical and research activities in the field of public health in health institutions on the territory of Serbia; and
- to give opinions on controversial ethical issues that are of importance for conducting scientific, medical and public health research in health institutions in Serbia etc.

According to the Law,[9] the faculties within universities where medical, dental and pharmaceutical sciences (faculties of health profession) are carried out, may conduct their activities through their organizational units if determined by the Ministry with a decision that they fulfil the necessary requirements for a certain type of healthcare institution, prescribed by the Law and regulations. The provisions of this Law and the regulations adopted for its implementation relating to a particular type of healthcare institution are applied accordingly, in accordance with the decision of the Ministry on the fulfilment of the prescribed conditions for performing healthcare activities.

The ethics committee in a healthcare institution

The Ethics Committee[10] is a professional body that monitors the provision and implementation of health care on the principles of professional ethics, the principles of respect for human values and the

[9] Law on Healthcare, Article 36.

[10] Law on Healthcare, Articles 130-132.

rights of the child, as well as codes of conduct for employees of the healthcare institutions.

The tasks of the ethics committee of the healthcare institution are the following:

1) to monitor and analyse the application of the principles of professional ethics in performing health care activities, as well as to propose measures for their improvement;
2) to monitor and analyse the implementation of codes of conduct for employees of the health care institution and propose measures for their improvement;
3) to approve the conduct of scientific research in the field of health, medical research, public health research, and monitor their implementation;
4) to approve the taking of human organs, cells and tissues from a living donor or a deceased person, in accordance with the law, and give an opinion on ethical and other issues in the process of transplantation, i.e. the application of cells and tissues;
5) to consider ethical issues and make decisions regarding the taking of parts of the human body for scientific-teaching purposes, in accordance with the law;
6) to consider ethical issues related to the implementation of measures for the treatment of infertility by biomedically assisted fertilization procedures, in accordance with the law;
7) to monitor, analyse and give opinions on the ethics of the relationship between health professionals, health care assistants and patients;
8) to monitor, analyse and give opinions on the application of the principles of professional ethics in the prevention, diagnosis, treatment, health care, rehabilitation, research, as well as on the introduction of new health technologies;
9) to contribute to the advancement of the application of the principles of professional ethics in the pursuit of the health care industry and the development of partnerships between health care professionals, healthcare associates and patients;
10) to perform an advisory function and consider other ethical issues in performing the activities of a health care institution; and
11) to cooperate with the ethical committee of the competent chamber.

The members of the ethics committee are appointed from among the health workers or associates employed in the health institution.

A person may be appointed as a member of the ethics committee who, in the pursuit of the profession, is distinguished in respecting the moral and ethical principles of the medical profession.

Using corpses for practical training at faculties

Another important aspect of respecting ethical principles in healthcare and higher education are the articles[11] regarding corpses and their use in practical teaching.

Faculties may take over the bodies, organs and tissues of deceased and identified persons for the purpose of practical training in the following cases:

1) if the deceased person has explicitly, in writing, bequeathed their body for the purpose of practicing;
2) if it is a deceased person without a family, and using the body has not been explicitly opposed in writing;
3) with the consent of the family, if the deceased person did not explicitly, in writing, oppose it.

The decision to take over the body is made by the ethics committee of the faculty. The faculty cannot use the body of a deceased person who is without a family for practical training for more than six months from the date of taking over.

The faculty is obliged to bury the body of the deceased at its own expense after completing the process of practical teaching in anatomy.

Conflict of interest

Conflict of interest is a situation in which a healthcare professional or a health associate has a private interest that influences, may influence, or may affect the impartiality and objectivity of the said professional in their performance of healthcare activities.

A Healthcare worker and health associate who perform healthcare activities in a publicly owned health institution, a member of bodies or

[11] Law on Healthcare, Articles 210-216.

professional bodies in a publicly owned health institution, as well as members of their immediate families, must neither seek nor receive money, gifts, services or any other benefits to themselves, to the members of their family or to persons who may reasonably be regarded as having an interest. This interest may affect the impartiality or professional performance of their duties, or may be considered a reward in the exercise of their duties, and performing healthcare activities.

The Ethics Committee of the health institution, in addition to the competencies referred to in Article 131 of this Law, is also responsible for:

1) providing advice and guidance on the prevention of conflicts of interest and corruption;
2) identifying and reducing the risk of conflict of interest and corruption;
3) training and raising awareness of employees on the prevention of conflicts of interest and corruption;
4) giving opinions in cases of suspected conflict of interest and corruption;
5) taking measures necessary for the protection of persons who provided information on cases of conflict of interest and corruption.

THE ETHICS COMMITTEE AT THE FACULTY OF MEDICINE FROM THE UNIVERSITY OF BELGRADE

The procedures of the Ethics Committee at the Faculty of Medicine are regulated by the Rulebook on the work of the ethics committee adopted by the Academic Council of the Faculty in 2017 and the Rulebook on the taking over and use of corpses for the purpose of carrying out practical classes at the Faculty of Medicine. These rulebooks are based on the Law on Higher Education, Basics for the Code of Academic Integrity in Higher Education Institutions in the Republic of Serbia,[12] the

[12] Basics for the Code of Academic Integrity in Higher Education Institutions in the Republic of Serbia (2016) National Council for Higher Education, available at http://nsvo.gov.rs/wp-content/uploads/2013/11/Osnove-za-kodeks-o-akademskom-integritetu.pdf (accessed January 20, 2021).

Code of conduct in scientific research,[13] rules and rulebooks of the University of Belgrade and the Law on Healthcare.

The Role of the Ethics Committee in higher education and science

The Rulebook on the work of the ethics committee regulates more closely the composition, jurisdiction, manner of operation and other issues of importance for the work of the Ethics Committee at the Faculty of Medicine from the University of Belgrade (hereinafter referred to as "the Ethics Committee") within the competence established by this Rulebook, the Code of Professional Ethics of the University of Belgrade[14], Rules of Procedure of Ethics Committee and the Professional Ethics Committees University of Belgrade[15] and the Rulebook on the Procedure for Determining Non-academic Behaviour in the Publication at the University of Belgrade.[16]

The Committee President, the members and their deputies are appointed by the Academic Council and proposed by the Dean, in consultation with the Departments.

The Ethics Committee consists of 7 members, 5 of whom are teachers and associates at the Faculty of Medicine; one member is a medical student, and one is a non-teaching staff employed by the Faculty of Medicine. Each member has their deputy.

The competencies of the Ethics Committee include the following:

1) Leading the procedure for determining violations of the Code of Professional Ethics University of Belgrade in accordance with the

---

[13] Code of conduct in scientific research (2018) National Council for Scientific and Technological Development, available at http://www.arh.bg.ac.rs/wp-content/uploads/dokumenti/Pravilnici/Kodeks-naucnoistrazivacki-rad-2018.pdf (accessed Jan. 20, 2021).

[14] Code of Professional Ethics of the University of Belgrade (University of Belgrade Publication 193/16), available at www.arh.bg.ac.rs/wp-content/uploads/dokumenti/Pravilnici/Kodeks-profesionalne-etike_2016.pdf (accessed January 20, 2021).

[15] Rules of Procedure of Ethics Committee and Professional Ethics Committees University of Belgrade, www.bg.ac.rs/files/sr/univerzitet/univ-propisi/Pravilnik-eticke-komisije-odbor.pdf (accessed January 20, 2021).

[16] Rulebook on the Procedure for Determining Non-academic Behaviour in the Publication at University of Belgrade (University of Belgrade Publication 193/16, 196/16,197/17, 199/17, 203/18, 206/18), www.bg.ac.rs/files/sr/univerzitet/univ-propisi/Pravilnik-neakademsko-pisanih-radova.pdf (accessed January 20, 2021).

Rules of Ethics Committees and of the Professional Ethics Committee of the University of Belgrade and the Rulebook on the process of determining non-academic behaviour in the publication at the University of Belgrade;

2) Considering all ethical issues related to scientific and research teaching work of teachers, associates and students of the Faculty of Medicine and giving the necessary approvals for the design of doctoral theses and other research papers;
3) Conceptualizing the rules, scope and content of work-related ethics principles at the Faculty of Medicine;
4) Analysing ethics in the practical work of teachers, associates and students at the Faculty of Medicine;
5) Providing remarks on the qualification and competence of researchers and analysing the potential risks that may arise during work experimental phases;
6) Taking care that respect for human beings is ensured in every scientific research, in particular the rights to human dignity;
7) Especially monitoring research work regarding children, pregnant women, mentally ill persons, prisoners and other special communities;
8) Promoting the principles of medical ethics both within teaching and educating health professionals, and in creating a habit for ethical behaviour in daily life and practice; and
9) If necessary, hiring experts from specific fields, such as organ transplants, AIDS and other related ethical consideration.

The Code of Professional Ethics of the University of Belgrade is adopted for preserving the dignity of the profession, the promotion of moral values, the protection of the value of knowledge and the awareness of all the members of the University of Belgrade. The Code is applied in professional and public action by all members of the university community: teaching and scientific staff, students, non-teaching staff, as well as persons who are not members of the academic community but participate in their work.

The procedures prescribed by the Code are not a substitute for civil, criminal, administrative, disciplinary and other proceedings, which are governed by the laws and by-laws of the University and its members.

Ethics Committee at the Faculty of Medicine resolves numerous cases regarding the ethical aspect of the scientific work of students, researchers and teaching staff. In these cases, it takes care about respecting basic principles in biomedical research such as: autonomy of the patient, its well-being, safety and justice.

The principle of autonomy is implemented in practice through a free and informed consent procedure that can be withdrawn at any time without negative consequences.[17] Well-being and safety summarize the moral obligation of providing maximum potential well-being, and at the same time providing minimum potential harm. Justice is particularly relevant to selecting research participants, since the selection criteria for participants must relate to the purpose of the research.

Other cases resolved by the Ethics Committee include non-ethical behaviour of students, teachers and associates such as plagiarism, false authorship, fabrication and falsification of results and auto-plagiarism. Additionally, several cases were devoted to resolving conflicts between students and mentors, or co-authors of publications.

The Role of the Ethics Committee in health care topics

As previously stated, according to the Law on Healthcare, the Faculty of Medicine carries out healthcare activities through their organizational units as a healthcare institution. That is the reason why the Ethics Committee has another role, to resolve issues in healthcare activities of the Faculty, according to the Law on healthcare.

In this role, the Ethics Committee has to supervise and control compliance with physician duties in performing their professional activity conscientiously, in accordance with the rules of the medical profession and with due care, in accordance with the principles of medical ethics and the principles of humanity. The physician should always keep in mind the absolute obligation to protect human life and health, while respecting the basic principle of medical ethics "do no harm."[18]

Additionally, it monitors and analyses the application of other principles of professional ethics in performing healthcare activities and

[17] Guide for research ethics committee members (2014) Council of Europe, https://www.coe.int/t/dg3/healthbioethic/activities/02_biomedical_research_en/Guide/Guide_EN.pdf (accessed January 20, 2021).

[18] Code of medical ethics of the Medical chamber of Serbia (2016) Official Gazette RS 104/16, Art. 4, www.lks.org.rs/Storage/Global/Documents/Akti/medicinska%20etika.pdf.

proposes measures for their improvement; approves the conduct of scientific research in the field of health, medical research, public health research, and monitors their implementation; approves the taking of human organs, cells and tissues from a living donor or a deceased person in accordance with the law; considers ethical issues and makes decisions regarding the use of parts of the human body for scientific-teaching purposes, in accordance with the law; monitors, analyses and gives opinions on the ethics of the relationship between health professionals, health care assistants and patients.

A particular role of the Ethics Committee, in this competence, is in using corpses for practical teaching. This role is regulated by the Rulebook on taking and using the bodies of the deceased for the purpose of carrying out practical classes at the Faculty of Medicine adopted at the Academic Council of the Faculty in 2018.

This Rulebook governs the manner and procedure of taking over and using the bodies of deceased and identified persons for the purpose of performing the practical classes in anatomy at the integrated academic studies of medicine, specialization studies, as well as in continuing education courses (hereinafter referred to as "practical training") at the Faculty of Medicine in Belgrade.

The Faculty may take cadavers for practical training in the following cases:

1) if the deceased person expressly, in writing, bequeathed their body for the purpose of practical training;
2) if it is a deceased person without a family, who not explicitly and in writing opposed to having their body used;
3) with the consent of the family, if the deceased person has not explicitly, in writing form, opposed it.

The Faculty can directly take over the cadavers for practical training, only if there is a death certificate signed by the doctor of medicine competent to determine death and provided there are no legally prescribed reasons for the mandatory autopsy.

The Faculty will not take over the body of a person who has died of an infectious disease, as well as the body of a person who underwent pronounced post-mortem changes that prevented fixation (embalming).

The decision to take over the cadaver is taken by the Ethics Committee.

The body of a deceased person who is without a family cannot be used by the Faculty in a practical training for more than six months from the date of takeover.

If a family member of a deceased person who was not aware of the fact at the time of death, within six months from the date of taking over the body by the Faculty, submits a written request for the recovery of the body of a deceased person, the Faculty is obliged to return the body of the deceased to the members of families, in accordance with the law.

The Faculty is obliged to keep all relevant information as a professional secret for the persons whose organs or parts of the body have been taken over for practical training and other necessary documentation of the deceased person whose body was taken.

The information includes the name and surname of the deceased, date of birth, place and date of death, cause of death, the number from medical records, the number of the plaque with the body of the deceased person, with place and date of the burial.

The documentation additionally includes: death certificate, excerpt from death certificates, identity card, health ID and testamentary statement bodies.

The data and documentation are permanently stored and placed by the Faculty for inspection by the relevant departments of the Faculty, the Ministry responsible for affairs of education, the Ministry responsible for internal affairs, and other competent authority, in accordance with the law.

The Faculty is obliged to treat the body of the deceased with dignity and to use it solely for the purpose of practicing.

The use of mobile phones and any kind of photography and recording is prohibited in the classroom where practical instruction is provided.

Clothes and footwear of teachers and associates, employees, students in the cadaver room have to be appropriate. The dress code and the code of conduct of the participants in practical teaching must be prominently displayed and accessible to all participants.

The Faculty is required to bury the body of the deceased after the completion of the classes, at its own expense.

The act of burial is announced in the form of a press release and a paid advertisement in the fund's public information; the funeral ceremony involves keeping an honour guard by the Faculty and students.

The Faculty is obliged, within the limits of its possibilities, to respect special wishes regarding the burial, cremation, religious ceremony and other expressed wishes of the testator regarding the treatment of their body for the performance of practical classes.

The Ethics Committee of the Faculty oversees the entire procedure of taking and using a cadaver and realizing practical teaching.

Continuing education courses using cadavers are conducted after obtaining the approval of the Ethics Committee. Courses using cadavers can be undertaken and organized at the Faculty, if they are accredited through the competent chamber and the Health Council of the Republic of Serbia, under specific conditions where the Faculty is the principal course organizer.

CONCLUSION

The Ethics Committee at the Faculty of Medicine from the University of Belgrade has very different and complex roles because of the large number of teaching staff, associates and students on numerous study programs, the necessity for the approval of ethics committee in different topics and phases of studies and finally, the different roles of the Faculty in higher education and the healthcare system.

According to favorable laws, the aims of the Ethics Committee include the following:

1) Preserving the integrity of higher education at the Faculty
2) The process of detecting non-academic behaviour
3) Imposition of the measures for established non-academic behaviour
4) Monitoring research work
5) Resolving conflicts between teachers and students
6) Supervising compliance with physician duties to perform their professional activity conscientiously
7) Approving the conduct of scientific research in the field of health, medical research, public health research and
8) Monitoring the use of cadavers in practical classes.

The new challenge for the Ethics Committee should include the possibility for consultation and guidance on bioethical issues that inevitably arise in the course of patient care and research, such as the possibility for ethics consultation at the Icahn School of Medicine at Mount Sinai Hospital, New York.[19] The ethic consultation in patient care topics could be requested by a healthcare professional including faculty staff, attending or consulting physicians, nurses, genetics counsellors, patient representatives from teaching hospitals, etc. When a difficult decision needs to be made about a case or a policy, the Ethics Committee could be available in order to help to:

- Clarify the ethical question at hand;
- Promote discussion of responsibilities and standards related to patient care;
- Encourage shared decision making among patients, families, and their healthcare team;
- Make recommendations to the individuals involved; and
- Provide moral support for a team in a difficult decision.

Research ethics consult service could help investigators to identify ethical issues in research proposals and to design their studies with the aim to meet the highest ethical standards, such as informed consent, research risks, inducements, vulnerable populations, conflict of interest, confidentiality, recruitment, assessment of decisional capacity and questions about surrogate consent.

In this role, the Ethics Committee could act in an educational form, while also preventing possible ethical offenses and responsibilities of health care professionals and investigators.

REFERENCES

Basics for the Code of Academic Integrity in Higher Education Institutions in the Republic of Serbia (2016). National Council for Higher Education. Available at http://nsvo.gov.rs/wp-content/uploads/2013/11/Osnove-za-kodeks-o-akademskom-integritetu.pdf. Accessed January 20, 2021.

Code of conduct in scientific research (2018). National Council for Scientific and Technological Development. Available at www.arh.bg.ac.rs/wp-

[19] Ethics Services – The Bioethics Program at the Icahn School of Medicine at Mount Sinai (ISMMS), https://icahn.mssm.edu/education/students/ethics (accessed Sept. 16, 2019).

content/uploads/dokumenti/Pravilnici/Kodeks-naucnoistrazivacki-rad-2018.pdf. Accessed Jan. 20, 2021.

Code of Professional Ethics of the University of Belgrade. University of Belgrade Publication 193/16. Available at www.arh.bg.ac.rs/wp-content/uploads/dokumenti/Pravilnici/Kodeks-profesionalne-etike_2016.pdf. Accessed January 20, 2021.

Code of medical ethics of the Medical chamber of Serbia (2016). Official Gazette RS 104/16, Art. 4. Available at www.lks.org.rs/Storage/Global/Documents/Akti/medicinska%20etika.pdf.

Guide for research ethics committee members (2014). Council of Europe. Available at www.coe.int/t/dg3/healthbioethic/activities/02_biomedical_research_en/Guide/Guide_EN.pdf. Accessed Jan. 20, 2021.

Ethics Services – The Bioethics Program at the Icahn School of Medicine at Mount Sinai (ISMMS). Available at https://icahn.mssm.edu/education/students/ethics. Accessed Sept. 16, 2019.

Law on Higher Education. Official Gazette RS 88/17, 27/18 - an. law, 73/18), Article 7. Available at: www.nat.rs/wp-content/uploads/2019/01/Zakon-o-visokom-obrazovanju-2017-ENGL.pdf. Accessed January 20, 2021.

Law on Copyright and Related Rights. Official Gazette RS 104/2009, 99/2011, 119/2012 and 29/2016. Available at www.zis.gov.rs/upload/THE%20LAW%20ON%20COPYRIGHT%20AND%20RELATED%20RIGHTS-version%20in%20force%204.1.2013._.pdf. Accessed January 20, 2021.

Law on Healthcare. Official Gazette RS 25/19. Article 155. Available at http://www.skriningsrbija.rs/files/File/English/Republic_of_Serbia_Healthcare_Law.pdf. Accessed January 20, 2021.

Rules of Procedure of Ethics Committee and Professional Ethics Committees University of Belgrade. Available at www.bg.ac.rs/files/sr/univerzitet/univ-propisi/Pravilnik-eticke-komisije-odbor.pdf. Accessed January 20, 2021.

Rulebook on the Procedure for Determining Non-academic Behaviour in the Publication at University of Belgrade. University of Belgrade Publication 193/16, 196/16,197/17, 199/17, 203/18, 206/18. Available at www.bg.ac.rs/files/sr/univerzitet/univ-propisi/Pravilnik-neakademsko-pisanih-radova.pdf. Accessed January 20, 2021.

SPORTS AND THE ETHICAL IMPLICATIONS OF PERFORMANCE ENHANCING SUBSTANCES

Levente Nagy [1]

INTRODUCTION

In sports, it is apparent the system itself compels athletes to exploit themselves. Competitiveness is everywhere, daily, as every person competes with other individuals in a way or another. There are just as many examples of competition in the economy. Sports and economy are similar in the sense that there is only a minimal difference between winners and losers. Nevertheless, these minor differences in achievement may lead to enormous differences in prestige. This is why each competitor is ready to do anything to be among the winners. Thus competitors in sports and the economy share several common attitudes and characteristics.

What should be done to enhance performance in sport while eliminating cheating? In certain sporting types, performance-enhancing drugs are relatively common. A solution for this issue seems to be abandoning the concept of leading sports. Legalizing sports achievements is the only way to a certain safe level. This should result in sport being pursued per se, without any winners.

According to the results of a 2011 survey,[2] over one-third of athletes had already used some kind of performance-enhancing drugs. However, drug tests were able to only reveal about 2% of these cases.

---

[1] Eötvös Loránd University, Budapest, Hungary.

[2] Tim Rohan, "Antidoping Agency Delays Publication of Research," *New York Times* (2013), https://www.nytimes.com/2013/08/23/sports/research-finds-wide-doping-study-withheld.html (accessed November 7, 2020).

Sports are two-fold. According to the rules, there is a legal and a moral side to sports. From the perspective of legality, fair play refers to respecting the rules. A fair player meets the rules and regulations. Legality thus refers to truthfulness with a type of authority or power in the background.

However, fair play is not based on emotions, as people do not have an identical sense of morality. Morality implies that people participate in sports games along with moral values. Participation and behaviour are not determined by any authority or power, but by moral principles sometimes enforced (even against ourselves). Fair sport can only appear on this level. As already seen, from a legal point of view, truthfulness is linked to external regulation. If that regulation does not operate – the judge does not see a problem, the electronic display system does not operate, the cameras do not record a decisive moment, the adversary is still judged in the same way, and thus no cheating is supposed to take place.

In Kantian thought, we are citizens of two worlds,[3] thus both sides of fair play – legality and morality – have to be considered at all times.

There are different understandings of sports, be they normative or descriptive. In a normative sense, sport is an activity and it meets the requirements of fairness. In a descriptive sense sport is everything that people call sports. Normatively, sports create values, while descriptively, sports may destroy existing values. Bullfighting cannot be considered a sport in a normative sense because death and destruction are not values, but, in a descriptive sense, it is still a type of sport because the fighter achieves their aim (victory) through struggle.

Sports philosophy is a branch of applied philosophy and it aims to investigate the true sense and ultimate goal of sport, i.e. the clarification of the meaning of sport. Sports philosophy relates its findings to reality. In order to assess right and wrong in sports norms, ideals and particular value systems are needed. The ideal image and state of sports must also be compared to the reality of sports. Several values which should be considered in this case include virtues, skills, physical and moral strength and struggle.

---

[3] Immanuel Kant, *Kritik der reinen Vernunft* (Budapest: Osiris, 2004), 106.

HISTORICAL OVERVIEW

Performance-enhancing drugs have been used since Antiquity when they were the instruments of druids, shamans or magicians. Therefore by tradition, the use of performance-enhancing substances goes back to the world of beliefs and religion.

In modern sports, doping was first related to cycling, a sport that became popular in the late 1800s. Doping made its first victim in bicycling in early 1900.[4]

Today the most common drugs used for doping purposes are a type of amphetamines. There are two groups of drugs, which can be identified. The first group, the classical drugs, enhances sport performance during the competitions themselves. However, as all athletes are tested today before sports events, a new group of drugs has become more common, which can be used during the training period instead.

Various forms of testing were introduced in the 1960s. The Olympic Games organized in Mexico City were the first Olympics where doping tests were used. It was a straightforward test and only three cases were identified.

In the 1970s and 1980s, doping tests were not systematically used and furthermore, many of them proved to be unreliable.[5] Tests identifying anabolic steroids, the most common performance-enhancing drugs, were first used in the 1970s. As a result of this new type of testing, eight athletes were disqualified from the Montreal Olympic Games of 1976. However, tests were still not efficient enough because the technology itself was underdeveloped and unreliable.

Today the situation is different. Analytical methods have improved, while the antidoping activities of national and international sport organizations have also become more active, despite that both antidoping tests and campaigns bear high costs.

Moller pointed out that the roots and the difficulties of antidoping movements remain unchanged because the notion of doping has never been adequately defined and clarified. Furthermore, in Moller's view,

---

[4] P. Dimeo, *A History of Drug Use in Sport: 1876–1976: Beyond Good and Evil* (Milton Park: Routledge, 2007).

[5] I. Waddington, *Sport, Health and Drugs* (Milton Park: Routledge, 2000).

because of the notion's vagueness, antidoping legal action has also remained inadequate.[6]

When an athlete craves winning, the use of 'mild' drugs may seem 'pardonable' even part of the game.

The social environment, especially the trainers and the members of the competitor's team, can have an enormous influence on the likelihood of unsportsmanlike behaviour. Thus the sports teams themselves, the athletes' social environment and the moral beliefs of the team members all have a powerful impact on the occurrence of doping.

ETHICAL PROBLEMS

The so-called elite amateurism has disappeared from sports competitions and, in Renson's view, sport has become an entertaining event instead.[7] Sport has become show-business-like, meaning that it is not the fairness of the game that matters, but the 'pure play.'

In Renson's view, the reality is that spectacular features in sport dominate over its competitiveness. This leads to the conclusion that one should not emphasize the concept of fair play because it is non-existent. Consequently, reality and morality can be mistakenly thought to be interchangeable. Only reality should be considered, as competing athletes do not need the idea of fairness; instead, they need adequate labour protection rules similar to employees in show business. In the absence of such protective measures leading athletes are defenceless and exposed to accidents, doping pressure or health risks. Thus sports is largely about money and it is not likely that this would change.

Sports are a value-creating activity. If sports is practiced with the sole financial aim, it would lead to estrangement. In a normative sense, only those activities which bring about activities can be considered a sport. In a descriptive sense, as already seen, sporting activities can also be destructive and still be called sports.

The aim of the veritable athlete is to become the best and not simply the winner. Any victory is good as long as it involves the improvement

---

[6] V. Møller, *The Ethics of Doping and Anti-doping. Redeeming the Soul of Sport?* (Milton Park: Routledge, 2010).

[7] R. Renson, "Fair play: Its origins and meanings in sport and society," *Kinesiology* 41 (2009).

factor, meaning that all the parties involved, including the opponent, could become better through as a result of it.

Due to sponsorships and professionalization, excellence in performance is highly significant in the lives of athletes. This may inevitably lead to temptation, which may lure athletes into choosing an illegal 'shortcut' instead of the burdensome road to victory. By choosing the easy road, athletes will not become the best in their branch of sport, but they may secure financial stability, fame and respect – at least until reality is revealed.

Doping is thus a 'shortcut' to an easy victory. It is also a signal that the current status of elite sport is not acceptable. Doping denotes a variety of substances that have varied effects on health. When investigating the illegal substances used to enhance sports performance, it is useful to divide them into two distinct groups, i.e. substances and techniques.

Tamburini's classifies substances as follow:

1. Harmful but legal substances, prescribed by a doctor. This group includes anabolic steroids, human growth hormones, beta blockers, various stimulants (amphetamines) etc. Prolonged and uncontrolled usage of these substances may lead to severe health risks. The side effects of short-term use and controlled dosage are less known.
2. Harmful and illegal substances, including central nervous system stimulants.
3. Harmless and legal substances including diuretics and caffeine. These substances are used for many purposes and there are no limits to determine when and how to use them.[8]

WHAT IS DOPING AND WHAT IS SPORT?

Defining doping is a difficult task as several definitions already exist and none of them seems to be sufficiently inclusive. Nevertheless, a description of the phenomenon must be presented because of the need to debate new trends in sports, as well as the need to oblige athletes to comply with the rules.

[8] C. Tamburini, T. Tännsjö, *Values in Sport* (Milton Park: Routledge, 2000).

Certain substances and techniques can be used for the purposes of doping. Blood doping is the best-known example. When correctly done, although there is no known health risk identified with it, it remains illegal. A widely discussed controversy is that the same effect can be reached by high-altitude training, which is a legal technique. This shows that the notion of doping can be rather ambiguous.

It must be highlighted that despite its prohibition, doping still exists. WADA (the World Anti-Doping Association) defines doping by listing illegal substances and a description of antidoping behavioural tests. Some of the behavioural requirements include the availability of athletes during the entire preparation period in order to take random drug tests.

WADA states that using an illegal substance or technique is a direct violation of doping rules. If athletes object to a doping test, or, if the test is positive, the athletes may be banned from sports activities for a certain period of time. Every year, WADA updates its list of illegal substances and techniques.

In short, doping can be defined as the abuse or misuse of performance-enhancing substances and techniques. Defining the notion of doping also implies defining the notion of sport. This a similarly difficult task, but at least a broad definition is needed in order to differentiate between sport- and non-sport activities.

Moller proposed four main criteria for a sport. (1) The activities must be performed within the framework of competition (2) they should not serve any external purpose, (3) they must be organized and done according to a certain set of rules, (4) athletes aim at success and excellence.[9]

Further on, McNamee identifies sports as ritual activities done in a well-regulated and competitive way. In his interpretation, any sport has an ethical dimension.[10]

Suits devoted his entire research career to defining sport. His views can be briefly summarized as follows:

1. A goal needs to be set up before the competition (score a goal, jump over a certain height etc.)
2. Rules need to be established in order to identify what is and is not

[9] V. Møller, *The Ethics of Doping and Anti-doping. Redeeming the Soul of Sport?* (Milton Park: Routledge, 2010).

[10] M. McNamee, *Sports, Vices, and Virtues* (Milton Park: Routledge, 2008).

permissible.[11]

Suits argued that the athletes who do not comply with these views do not, in fact, pursue a sport. Furthermore, Suits offered a classical definition of sport: 'a voluntary attempt to overcome unnecessary obstacles.'[12]

WHAT IS THE PROBLEM WITH DOPING?

When arguing against doping, there are three main issues to consider.

1. Doping influences health in a negative way. However, pro-doping arguments against this issue are that making these substances illegal, which deprives athletes of their rights to decide freely on what to do with their own bodies.
2. Doping has harmful effects on society as substances are sometimes used by amateurs and young people.
3. Doping needs to be blocked on moral grounds. Temporarily it can be to the advantage of certain athletes while being at the disadvantage of non-users who may unjustly be deprived of the possibility to win. This situation contradicts one of the main principles of sport, that of competition, as the results are determined by external and artificial factors instead of the athletes' capabilities and skills. Also, sports, in general, hold a certain factor of uncertainty and may generate joy and excitement in supporters and spectators. Doping would ruin the uncertainty factor, as the legalization of various doping substances and techniques would contribute to the pre-calculability of results, thus significantly weaken the inherent joy and excitement in competitive sport.

As already argued before, the financial gains of doping are enormous. Doping does not belong to the sphere of excellence in sport, but rather to the realm of temporary financial benefits and false fame.

WHY ARE DRUGS USED IN SPORTS?

McNamee presents several characteristics of doping: (1) performance enhancement (2) expansion of training time and enhancement of training

[11] B. Suits, "Tricky Triad: Games, Play and Sport," *Journal of Philosophy of Sport* 13 (1988).
[12] B. Suits, *The Grasshopper: Games, Life and Utopia* (Peterborough: Broadview Press, 2005).

efficiency; (3) unnatural; (4) addictive; (5) harmful for health; (6) unfair advantage over others.[13]

Instead of doping, McNamee recommends other performance-enhancing techniques, including the introduction of refined and special training, the use of hypoxic environments and tents, dieting awareness, improved technical conditions and training. In his opinion, all these conditions may contribute to enhanced performance without the use of drugs.

McNamee also adds that medicine is crucial for healing purposes but is unacceptable inin sports and competition. Further, he added cynically that drugs might represent the price spectators and fellow athletes pay for the progressive nature of medical science.

Furthermore, McCalla adds that it is generally thought that the doping problem cannot be solved by making drugs more accessible.[14]

ARGUMENTS AGAINST DOPING

The purpose of the antidoping movement is to save sports. By definition, doping is using stimulating agents intentionally, with the aim of artificial or unfair enhancement of performance. But how can intentionality be proved? Athletes may defend themselves by arguing that they have only taken prescribed drugs, designed to improve their health. This argument is already known to be frequently used.

One of the essential antidoping arguments is its unnaturalness. There are two important questions related to this issue. The first issue is centred around what is considered natural versus unnatural, while the second one explores the arguments against the unnatural aspect of doping. There are no clear boundaries between what is natural and what is not. So, are 'natural' performance-enhancing substances acceptable and the others are automatically dismissed? In addition, sport itself can be seen as a sequence of unnatural activities performed in unnatural circumstances. Furthermore: why would the unnatural automatically be deemed immoral? Is the unnatural body considered ideal, whose functioning is not supported by normal human biological functions?

---

[13] M. McNamee, *Sports, Vices, and Virtues* (Milton Park: Routledge, 2008).
[14] S. S. McCalla, N. Shepherd, "Moral/Ethical Analysis of Performance Enhancement in Sports," *International Journal of Arts & Sciences* (2014), www.universitypublications.net/ijas/0704/pdf/T4N233.pdf.

What natural when using performance-enhancing drugs and techniques? Sleeping in an oxygen tent to stimulate red blood cells to multiply and improve the body's oxygen supply seems to be acceptable because it is the body itself which produces these substances.

The biological antidoping argument thus asserts that doping is wrong because the substances used are harmful to the human body. However, not all illegal substances are harmful, so other arguments need to be found to declare doping harmful.

What should be the boundaries of controlled interference? How could doping be prevented? In addition to regular medical tests, athletes should be presented with the ethical side of the sport. The most common ethical argument is that doping allows using physical advantages in an unfair way. We may argue that there already is a considerable degree of inequality in sport. Then why should only doping be blamed for uneven conditions and unequal opportunities?

A further problem is that the same substances may have a different impact on different individuals. Considering this, legalizing certain forms of doping would not automatically lead to equality in sport.

PRO-DOPING ARGUMENTS

Tamburini believes that the prohibition of doping should be repealed because it is unreasonable[15]. Two major problems related to doping can be identified. The first problem relates to physical (medical) issues, while the other one is an ethical problem because doping does not comply with the principle of equal opportunity.

Tamborini's answer to the medical problem is that certain types of sport, including mountaineering, boxing or American football, can be more dangerous and can thus cause more severe injuries than doping itself (or might even lead to death). Furthermore, doping damages the health of athletes who use it, but it may also have effects on those who are reluctant to use drugs. Peer pressure might push these people into using drugs. Children and young people might also be affected, especially as successful athletes serve as role models, therefore young people might follow their examples.

Doping in professional sport is an increasingly severe issue. However, antidoping programs have been neglecting the ethical side of sports,

[15] C. Tamburini, T. Tännsjö, *Values in Sport* (Milton Park: Routledge, 2000).

although the issues inherent in sport ethics are of great significance for young generations; furthermore, they have pedagogical implications.

It is not right to curtail the professional freedom of athletes, thus smoking or the consumption of coffee is not prohibited.

A significant number of professional athletes are adolescents. The use of performance-enhancing drugs must be prohibited because they are not considered mature to make such decisions independently and responsibly. However, this way, they would be disadvantaged by the athletes who do use performance-enhancing drugs. Often the adults decide instead of the young and these adult decisions are often based on the idea of moneymaking and profit.

Another argument for doping is related to equal opportunity. There is no equal opportunity in sport, but the use of performance-enhancing drugs could counterbalance the differences. Nevertheless, people's reactions to drugs are not alike. Instead of talent and skills, technology would play the most decisive role in sports achievements. Doping would transform sport into a technological game; consequently, the winner would be predictable, excluding the surprise factor.

Doping is typical in scalarly measurable types of sport, in which the results can be measured in metres, centimetres, kilograms, or seconds. Creativity-based sports are more dependable on talent.

Today large companies are involved in sports organizations and they sponsor athletes who are thus advantaged as compared to unsponsored athletes. This is another factor of inequality. This leads to another pro-doping argument (although a weaker one) everybody must run the race, with or without drugs, while the same rule applies to each participant – this is one of the basic principles in sport.

The essence of sports is testing skills and capabilities. When using performance-enhancing drugs, it is not the athlete but the substance which contributes to victory. Doping implies minor yet decisive differences.

Simon argues against doping.[16] In his opinion, eating habits and training methods should be updated and modernized, thus contributing to the enhancement of performance in sport.

Furthermore, if doping is legalized, athletes would not be considered humans and characteristic features such as courage, perseverance and motivation would not improve.

---

[16] R. L. Simon, "Good Competition and Drug-Enhanced Performance," in *Philosophic Inquiry in Sport*, ed. William J. Morgan, Klaus V. Meier (Champaigne, 1988).

There are several other arguments for the use of doping. Prohibition refers to the restriction of one's personal freedom by not making decision-making possible. Thus the prisoner`s dilemma appears.

If nobody used doping, everyone had the chance to win the game, thus it is not in the interest of athletes to use drugs. On the other hand, if all participants use performance-enhancing drugs, then once again, everybody has the same chance of winning. If everybody uses the same performance-enhancing drug, but, due to side effects, it is an improper balance.

The problem is that doping implies the violation of rules, as it is prohibited. In this context, Tamburini believes that doping may be used in professional sport but it must be prohibited in amateur and recreational sport.

The professional athlete is driven by the desire to become number one and reach external goals such as money and prestige. Thus amateur and recreational sports have different aims. Those who play sports as amateurs have different, often internal goals.

Morgan's view on performance-enhancing drugs suggests that despite their harmful, addictive and morally destructive effects - they need to be legalized because there are too many chemicals, dietary supplements, painkillers in special diets and the dividing line between these substances and the performance-enhancing substances is very obscure. Furthermore, in addition to drugs, there are too many different unethical training methods, which are primarily undetectable and harmful.[17]

Competitors have varied access to resources such as trainers, psychologists, training equipment etc., resulting in unequal competition conditions. Still, modern sport is often identified with equal opportunity and fair play, and several experts say that the spread of doping would n end sports activities.

In addition, modern sport faces other problems, such as its connection to economy and politics. This might also contribute to the risk of doping.

17 W. J. Morgan, "Athletic Perfection, Performance-Enhancing Drugs, and the Treatment-Enhancement Distinction," *Journal of the Philosophy of Sport* 36(2) (2009), www.tandfonline.com/doi/abs/10.1080/00948705.2009.9714755.

The general opinion about doping used to be that it is acceptable if it does not harm the athletes' health. Thus medical and ethical arguments supplement each other.[18]

CONCLUSION

The antidoping arguments can be summarized as follows:

1. Competitive sports have a significant educational role in other areas of life as well.
2. From a medical point of view, doping in sport is incompatible with health.
3. From an ethical point of view, doping is incompatible with sports as athletes may be unfairly given advantages, thus damaging the sporting spirit.

According to Kant, the human being is the citizen of the Empire of Freedom, but also of the physical world of duties, and has to be present in both worlds. But can we exist in two worlds at the same time? Kant's philosophy gives us an answer to this question: the world of freedom does not exist without the world of duties. We are aware of our freedom only if we are aware of our duties. If we do not obey, then we succumb to our instincts and desires.

The ultimate goal in sport is victory, although not by all means. Victory is valuable only if it comprises true excellence, i.e. it also makes the adversary improve. If the adversary is defeated through cheating, there is no improvement; consequently, there is no victory. In this regard, the author of this essay agrees with Robert Simon, that all athletes must believe that victory takes place only if all participants improve. Values need to become universal in sport, independent of gender, age, or health. This needs to be the central idea of sports.

Victory in itself does not grant anyone supremacy over adversaries. The best sportsman does not always win, but a true sportsman is always determined to become the best.

My approach describes the issue of primacy in fairness. There are no ceremonies organized before competitions to make solemn promises. Participation is the promise itself.

18 P. Dimeo, *A History of Drug Use in Sport: 1876–1976: Beyond Good and Evil* (Milton Park: Routledge, 2007).

REFERENCES

Dimeo, P. *A History of Drug Use in Sport: 1876–1976: Beyond Good and Evil.* Milton Park: Routledge, 2007.

Kant, Immanuel. *Kritik der reinen Vernunft.* Budapest: Osiris, 2004.

McCalla, S. S., and N. Shepherd. "Moral/Ethical Analysis of Performance Enhancement in Sports." *International Journal of Arts & Sciences* (2014). www.universitypublications.net/ijas/0704/pdf/T4N 233.pdf.

McNamee, M. *Sports, Vices, and Virtues.* Milton Park: Routledge, 2008.

Møller, V. *The Ethics of Doping and Anti-doping. Redeeming the Soul of Sport?* Milton Park: Routledge, 2010.

Morgan, W. J. "Athletic Perfection, Performance-Enhancing Drugs, and the Treatment-Enhancement Distinction." *Journal of the Philosophy of* Sport 36(2) (2009). www.tandfonline.com/doi/abs/10.1080/0094 8705.2009.9714755.

Renson, R. "Fair play: Its origins and meanings in sport and society." *Kinesiology* 41 (2009).

Rohan, Tim. "Antidoping Agency Delays Publication of Research". *New York Times.* 2013. www.nytimes.com/2013/08/23/sports/research-finds-wide-doping-study-withheld.html.

Simon, R. L. "Good Competition and Drug-Enhanced Performance." In *Philosophic Inquiry in Sport. Human Kinetics*, ed. William J. Morgan, Klaus V. Meier. Champaigne, 1988.

Suits, B. "Tricky Triad: Games, Play and Sport." *Journal of Philosophy of Sport* 13 (1988).

——. *The Grasshopper: Games, Life and Utopia.* Peterborough: Broadview Press, 2005.

Tamburini, C., and T. Tännsjö. *Values in Sport.* Milton Park: Routledge, 2000.

Waddington, I. *Sport, Health and Drugs.* Milton Park: Routledge, 2000.

FSC
www.fsc.org
MIX
Papier | Fördert gute Waldnutzung
FSC® C083411

Kolibri 360 GmbH
Ferdinand-Jühlke-Straße 7
99095 Erfurt, Deutschland
produktsicherheit@kolibri360.de